Advanced Regression Models with SAS and R

California State University,
Long Beach

Advanced Regression Models with SAS and R

California State University,
Long Beach

Olga Korosteleva

CRC Press
Taylor & Francis Group
Boca Raton London New York

CRC Press is an imprint of the
Taylor & Francis Group, an **informa** business

CRC Press
Taylor & Francis Group
6000 Broken Sound Parkway NW, Suite 300
Boca Raton, FL 33487-2742

First issued in paperback 2020

© 2019 by Taylor & Francis Group, LLC
CRC Press is an imprint of Taylor & Francis Group, an Informa business

No claim to original U.S. Government works

Version Date: 20181101

ISBN 13: 978-0-367-73242-4 (pbk)
ISBN 13: 978-1-138-04901-7 (hbk)

Library of Congress Cataloging-in-Publication Data

Names: Korosteleva, Olga,
author.
Title: Advanced regression models with SAS and R / Olga Korosteleva.
Description: Boca Raton: Taylor & Francis, 2019.
Identifiers: LCCN 2018034122 | ISBN 9781138049017 (hardback)
Subjects: LCSH: Regression analysis--Textbooks. | SAS (Computer file) | R
(Computer program language)
Classification: LCC QA278.2.K6755 2019 | DDC 519.5/36--dc23
LC record available at https://lccn.loc.gov/2018034122

**Visit the Taylor & Francis Web site at
http://www.taylorandfrancis.com**

**and the CRC Press Web site at
http://www.crcpress.com**

Contents

Preface   xiii

1 Introduction: General and Generalized Linear Regression Models   1

1.1 Definition of General Linear Regression Model   1
1.2 Definition of Generalized Linear Regression Model   2
1.3 Parameter Estimation and Significance Test for Coefficients . . .   3
1.4 Fitted Model   3
1.5 Interpretation of Estimated Regression Coefficients   4
1.6 Model Goodness-of-Fit Check   4
1.7 Predicted Response   6
1.8 SAS Implementation   6
1.9 R Implementation   8
1.10 Example   9
    Exercises 15

2 Regression Models for Response with Right-skewed Distribution   23

2.1 Box-Cox Power Transformation 23
    2.1.1 Model Definition 23
    2.1.2 Fitted Model 24
    2.1.3 Interpretation of Estimated Regression Coefficients . . . 24
    2.1.4 Predicted Response 25
    2.1.5 SAS Implementation 25
    2.1.6 R Implementation 26
    2.1.7 Example 27
2.2 Gamma Regression Model 37
    2.2.1 Model Definition 37
    2.2.2 Fitted Model 37
    2.2.3 Interpretation of Estimated Regression Coefficients . . . 37
    2.2.4 Predicted Response 38
    2.2.5 SAS Implementation 38

	2.2.6	R Implementation	38
	2.2.7	Example	38
		Exercises	42

3 Regression Models for Binary Response   **49**

	3.1	Binary Logistic Regression Model	49
	3.1.1	Model Definition	49
	3.1.2	Fitted Model	50
	3.1.3	Interpretation of Estimated Regression Coefficients . . .	51
	3.1.4	Predicted Probability	52
	3.1.5	SAS Implementation	52
	3.1.6	R Implementation	52
	3.1.7	Example	52
	3.2	Probit Model	56
	3.2.1	Model Definition	56
	3.2.2	Fitted Model	57
	3.2.3	Interpretation of Estimated Regression Coefficients . . .	57
	3.2.4	Predicted Probability	57
	3.2.5	SAS Implementation	57
	3.2.6	R Implementation	58
	3.2.7	Example	58
	3.3	Complementary Log-Log Model	61
	3.3.1	Model Definition	61
	3.3.2	Fitted Model	61
	3.3.3	Interpretation of Estimated Regression Coefficients . . .	62
	3.3.4	Predicted Probability	62
	3.3.5	SAS Implementation	62
	3.3.6	R Implementation	63
	3.3.7	Example	63
		Exercises	66

4 Regression Models for Categorical Response   **71**

	4.1	Cumulative Logit Model	72
	4.1.1	Model Definition	72
	4.1.2	Fitted Model	73
	4.1.3	Interpretation of Estimated Regression Coefficients . . .	73
	4.1.4	Predicted Probabilities	74
	4.1.5	SAS Implementation	74
	4.1.6	R Implementation	74
	4.1.7	Example	75
	4.2	Cumulative Probit Model	80
	4.2.1	Model Definition	80

	4.2.2	Fitted Model	81
	4.2.3	Interpretation of Estimated Regression Coefficients . . .	81
	4.2.4	Predicted Probabilities	81
	4.2.5	SAS Implementation	82
	4.2.6	R Implementation	82
	4.2.7	Example	82
4.3	Cumulative Complementary Log-Log Model	86
	4.3.1	Model Definition	86
	4.3.2	Fitted Model	87
	4.3.3	Interpretation of Estimated Regression Coefficients . . .	87
	4.3.4	Predicted Probabilities	87
	4.3.5	SAS Implementation	88
	4.3.6	R Implementation	88
	4.3.7	Example	88
4.4	Generalized Logit Model for Nominal Response	92
	4.4.1	Model Definition	92
	4.4.2	Fitted Model	93
	4.4.3	Interpretation of Estimated Regression Coefficients . . .	93
	4.4.4	Predicted Probabilities	94
	4.4.5	SAS Implementation	94
	4.4.6	R Implementation	95
	4.4.7	Example	95
	Exercises	99

5 Regression Models for Count Response   **105**

5.1	Poisson Regression Model	105
	5.1.1	Model Definition	105
	5.1.2	Fitted Model	106
	5.1.3	Interpretation of Estimated Regression Coefficients . . .	106
	5.1.4	Predicted Response	107
	5.1.5	SAS Implementation	107
	5.1.6	R Implementation	107
	5.1.7	Example	107
5.2	Zero-truncated Poisson Regression Model	110
	5.2.1	Model Definition	110
	5.2.2	Fitted Model	111
	5.2.3	Interpretation of Estimated Regression Coefficients . . .	111
	5.2.4	Predicted Response	111
	5.2.5	SAS Implementation	111
	5.2.6	R Implementation	112
	5.2.7	Example	112
5.3	Zero-inflated Poisson Regression Model	115

		5.3.1	Model Definition	115
		5.3.2	Fitted Model	117
		5.3.3	Interpretation of Estimated Regression Coefficients	117
		5.3.4	Predicted Response	117
		5.3.5	SAS Implementation	117
		5.3.6	R Implementation	118
		5.3.7	Example	118
	5.4		Hurdle Poisson Regression Model	122
		5.4.1	Model Definition	122
		5.4.2	Fitted Model	122
		5.4.3	Interpretation of Estimated Regression Coefficients	123
		5.4.4	Predicted Response	123
		5.4.5	SAS Implementation	123
		5.4.6	R Implementation	124
		5.4.7	Example	124
			Exercises	127
6	**Regression Models for Overdispersed Count Response**			**137**
	6.1		Negative Binomial Regression Model	137
		6.1.1	Model Definition	137
		6.1.2	Fitted Model	138
		6.1.3	Interpretation of Estimated Regression Coefficients	138
		6.1.4	Predicted Response	138
		6.1.5	SAS Implementation	138
		6.1.6	R Implementation	139
		6.1.7	Example	139
	6.2		Zero-truncated Negative Binomial Regression Model	142
		6.2.1	Model Definition	142
		6.2.2	Fitted Model	143
		6.2.3	Interpretation of Estimated Regression Coefficients	143
		6.2.4	Predicted Response	143
		6.2.5	SAS Implementation	143
		6.2.6	R Implementation	143
		6.2.7	Example	144
	6.3		Zero-inflated Negative Binomial Regression Model	147
		6.3.1	Model Definition	147
		6.3.2	Fitted Model	148
		6.3.3	Interpretation of Estimated Regression Coefficients	148
		6.3.4	Predicted Response	148
		6.3.5	SAS Implementation	148
		6.3.6	R Implementation	149
		6.3.7	Example	149

6.4 Hurdle Negative Binomial Regression Model 153
    6.4.1 Model Definition 153
    6.4.2 Fitted Model 153
    6.4.3 Interpretation of Estimated Regression Coefficients . . . 153
    6.4.4 Predicted Response 154
    6.4.5 SAS Implementation 154
    6.4.6 R Implementation 155
    6.4.7 Example 155
    Exercises 159

7 Regression Models for Proportion Response   **167**
7.1 Beta Regression Model 167
    7.1.1 Model Definition 167
    7.1.2 Fitted Model 168
    7.1.3 Interpretation of Estimated Regression Coefficients . . . 168
    7.1.4 Predicted Response 168
    7.1.5 SAS Implementation 169
    7.1.6 R Implementation 169
    7.1.7 Example 169
7.2 Zero-inflated Beta Regression Model 173
    7.2.1 Model Definition 173
    7.2.2 Fitted Model 174
    7.2.3 Interpretation of Estimated Regression Coefficients . . . 174
    7.2.4 Predicted Response 174
    7.2.5 SAS Implementation 175
    7.2.6 R Implementation 176
    7.2.7 Example 176
7.3 One-inflated Beta Regression Model 181
    7.3.1 Model Definition 181
    7.3.2 Fitted Model 182
    7.3.3 Interpretation of Estimated Regression Coefficients . . . 182
    7.3.4 Predicted Response 183
    7.3.5 SAS Implementation 183
    7.3.6 R Implementation 183
    7.3.7 Example 184
7.4 Zero-one-inflated Beta Regression Model 189
    7.4.1 Model Definition 189
    7.4.2 Fitted Model 189
    7.4.3 Interpretation of Estimated Regression Coefficients . . . 190
    7.4.4 Predicted Response 190
    7.4.5 SAS Implementation 191
    7.4.6 R Implementation 191

7.4.7   Example 192

        Exercises 197

8   General Linear Regression Models for Repeated Measures Data   209

8.1   Random Slope and Intercept Model 209

    8.1.1   Model Definition 209

    8.1.2   Fitted Model 210

    8.1.3   Interpretation of Estimated Regression Coefficients   . . . 210

    8.1.4   Model Goodness-of-Fit Check 211

    8.1.5   Predicted Response 211

    8.1.6   SAS Implementation 211

    8.1.7   R Implementation 212

    8.1.8   Example 214

8.2   Random Slope and Intercept Model with Covariance Structure for Error 220

    8.2.1   Model Definition 220

    8.2.2   Fitted Model, Interpretation of Estimated Regression Coefficients, and Predicted Response 222

    8.2.3   Model Goodness-of-fit Check 222

    8.2.4   SAS Implementation 223

    8.2.5   R Implementation 223

    8.2.6   Example 224

8.3   Generalized Estimating Equations Model 237

    8.3.1   Model Definition 237

    8.3.2   Fitted Model, Interpretation of Estimated Regression Coefficients, and Predicted Response 239

    8.3.3   Model Goodness-of-Fit Check 239

    8.3.4   SAS Implementation 240

    8.3.5   R Implementation 240

    8.3.6   Example 241

    Exercises 248

9   Generalized Linear Regression Models for Repeated Measures Data   257

9.1   Generalized Random Slope and Intercept Model 257

    9.1.1   Model Definition 257

    9.1.2   Fitted Model, Interpretation of Estimated Regression Coefficients, Model Goodness-of-Fit Check, and Predicted Response 258

    9.1.3   SAS Implementation 258

    9.1.4   R Implementation 259

9.1.5   Example 260
9.2   Generalized Estimating Equations Model 265
9.2.1   Model Definition 265
9.2.2   SAS Implementation 265
9.2.3   R Implementation 265
9.2.4   Example 266
Exercises 271

10 Hierarchical Regression Model   **281**
10.1   Hierarchical Regression Model for Normal Response 281
10.1.1   Model Definition 281
10.1.2   Fitted Model, Interpretation of Estimated Regression
Coefficients, Model Goodness-of-Fit Check, Predicted
Response   282
10.1.3   SAS Implementation 282
10.1.4   R Implementation 283
10.1.5   Example 283
10.2   Hierarchical Regression Model for Non-normal Response   289
10.2.1   Model Definition 289
10.2.2   Fitted Model 290
10.2.3   Interpretation of Estimated Regression Coefficients   . . . 290
10.2.4   Model Goodness-of-Fit Check 290
10.2.5   Predicted Response 290
10.2.6   SAS Implementation 290
10.2.7   R Implementation 291
10.2.8   Example 291
Exercises 295

Recommended Books   **303**

List of Notation   **305**

Subject Index   **307**

Preface

This book presents the material that I had the privilege of teaching to Master's level students at California State University, Long Beach. The material was very well received by the students, and helped them a great deal in finding good jobs. Now, this book will serve as a textbook for an introductory upper-division undergraduate course in linear regression models.

Chapters 1 through 7 present 22 regression models: for responses with normal, gamma, binary, multinomial, Poisson, negative binomial, and beta distributions. The last three chapters (Chapters 8 – 10) deal with models for repeated measures and clustered data. Such aspects of regression are covered in this book as model setting and definition, formula for fitted model, goodness of model fit, interpretation of estimated regression parameters, and use of the fitted model for prediction. One example with complete SAS and R codes and relevant outputs is shown for each regression. Results in each example are discussed for SAS outputs, while R scripts and outputs are given without discussion.

The topic that was deliberately left out (not an easy decision) is model building via variable selection procedures, which would require significantly larger data sets and longer codes.

The settings for examples came from consulting projects, which I had been involved in for the past three years as the Faculty Director of the Statistical Consulting Group at my university. To simplify the discussion and focus on the models and their applications, the data have been "smoothed out" in a sense that the issues of missing values, outliers, multi-collinearity, and transformations of predictor variables don't have to be addressed.

The publisher's web site that accompanies this book <https://www.crcpress.com/Advanced-Regression-Models-with-SAS-and-R/Korosteleva/p/book/9781138049017> contains complete SAS and R codes for all examples, and data sets for all exercises in .csv format. A complete solutions manual is also available to instructors on the same web site.

Respectfully,
The author.

Chapter 1

Introduction: General and Generalized Linear Regression Models

1.1 Definition of General Linear Regression Model

Suppose a random sample of size n is drawn from a population, and measurements $(x_{i1}, x_{i2}, \ldots, x_{ik}, y_i)$, $i = 1, \ldots, n$, are obtained on the n individuals. The random variables $x_1, \ldots, x_k$ (with the lower-case notation for simplicity) are commonly termed *predictor variables* (or, simply, *predictors*), but, depending on the field of application, they may also be called *independent variables, covariates, regressors,* or *explanatory variables*. The y variable is the *response variable* (or, simply, *response*). Other terms include *dependent variable, variate,* or *outcome variable*. The *general linear regression model*[1] represents a relation between the response variable y and the predictor variables $x_1, \ldots, x_k$ of the form:

$$y = \beta_0 + \beta_1 x_1 + \cdots + \beta_k x_k + \varepsilon \tag{1.1}$$

where $\beta_0, \ldots, \beta_k$ are constant *regression coefficients*, and ε is a *random error* that has a normal distribution with the mean zero and a constant variance σ^2. Also, the random errors are assumed independent for different individuals in the sample. The parameters of the model $\beta_0, \ldots, \beta_k$, and the variance σ^2 are unknown and have to be estimated from the data. The relation between y

---

[1]The first rigorous treatment of linear regression was published in Pearson, Karl (1896). "Mathematical Contributions to the Theory of Evolution. III. Regression, Heredity, and Panmixia". *Philosophical Transactions of Royal Society of London, Series A*, 187, 253 – 318.

and $x_1, \ldots, x_k$ is not necessarily linear in the x variables as polynomial or interaction terms may be included, but it is necessarily linear in the beta coefficients.

Note that in the general linear regression model, the response variable y has a normal distribution with the mean

$$\mathbb{E}(y) = \beta_0 + \beta_1 x_1 + \cdots + \beta_k x_k, \tag{1.2}$$

and variance σ^2. Moreover, the values of y for different individuals are assumed independent.

1.2 Definition of Generalized Linear Regression Model

In a *generalized linear regression model*[2], the response variable y is assumed to follow a probability distribution from the exponential family of distributions, that is, the probability density function (in a continuous case) or probability mass function (in a discrete case) of y has the form

$$f(y; \theta, \phi) = \exp\left\{ \frac{y\theta - c(\theta)}{\phi} + h(y, \phi) \right\} \tag{1.3}$$

where $c(\cdot)$ and $h(\cdot)$ are some functions, and θ and ϕ are constants. The parameter θ is called *location parameter*, whereas ϕ is termed *dispersion* or *scale* parameter. Further, in a generalized linear regression model, the mean response $\mathbb{E}(y)$ is related to the linear combination of predictors $x_1, \ldots, x_k$ through an invertible *link function* $g(\cdot)$, that is, for some regression coefficients $\beta_0, \ldots, \beta_k$,

$$g\big(\mathbb{E}(y)\big) = \beta_0 + \beta_1 x_1 + \cdots + \beta_k x_k, \tag{1.4}$$

or, alternatively,

$$\mathbb{E}(y) = g^{-1}\big(\beta_0 + \beta_1 x_1 + \cdots + \beta_k x_k\big). \tag{1.5}$$

Note that the general linear regression defined by (1.1) is a special case of a generalized linear regression, since the normal density with mean μ and variance σ^2 belongs to the exponential family of distributions with $\theta = \mu$ and $\phi = \sigma^2$ (see Exercise 1.1). In addition, $\mathbb{E}(y) = \beta_0 + \beta_1 x_1 + \cdots + \beta_k x_k$, which tells us that the link function $g(\cdot)$ is the identity.

---

[2]Introduced in Nelder, J. and R. Wedderburn (1972). "Generalized linear models". *Journal of the Royal Statistical Society, Series A*, 135 (3): 370 – 384.

1.3 Parameter Estimation and Significance Test for Coefficients

In a generalized linear regression model, the regression coefficients $\beta_0, \ldots, \beta_k$, and the other parameters of the distribution are estimated by the method of the maximum likelihood estimation. The estimates are computed numerically via an iterative process.

From the theory it is known that for a large number of observations, the maximum likelihood estimators have an approximately normal distribution. In particular, it means that a standard z-test is appropriate to use to test for equality to zero of each regression coefficient. For this test, the null hypothesis is that the coefficient is equal to zero, while the alternative is that it is not equal to zero. A p-value below 0.05 would imply that the regression coefficient is a significant predictor of the response variable at the 5% level of significance. Alternatively, this test may be conducted based on t-distribution or a chi-squared distribution.

1.4 Fitted Model

In accordance to (1.2), the *fitted* general linear regression model has the estimated (also called *fitted*) mean response

$$\widehat{\mathbb{E}}(y) = \widehat{\beta}_0 + \widehat{\beta}_1 \, x_1 + \cdots + \widehat{\beta}_k \, x_k, \tag{1.6}$$

and the estimated standard deviation $\widehat{\sigma}$. As explained above, $\widehat{\beta}_0, \ldots, \widehat{\beta}_k$ and $\widehat{\sigma}$ are the maximum likelihood estimates.

In view of (1.4) and (1.5), in the case of a generalized linear regression with the link function $g(\cdot)$, the fitted model has the estimated mean response $\widehat{\mathbb{E}}(y)$ that satisfies

$$g\big(\widehat{\mathbb{E}}(y)\big) = \widehat{\beta}_0 + \widehat{\beta}_1 \, x_1 + \cdots + \widehat{\beta}_k \, x_k, \tag{1.7}$$

or, equivalently,

$$\widehat{\mathbb{E}}(y) = g^{-1}\big(\widehat{\beta}_0 + \widehat{\beta}_1 \, x_1 + \cdots + \widehat{\beta}_k \, x_k\big) \tag{1.8}$$

where the estimates of the beta coefficients as well as the estimates for all the other parameters of the underlying distribution are obtained by the method of maximum likelihood.

1.5 Interpretation of Estimated Regression Coefficients

In a fitted general linear regression model with the estimated mean response defined by (1.6), the estimates of the regression coefficients $\widehat{\beta}_1, \ldots, \widehat{\beta}_k$ yield the following interpretation:

- If a predictor variable x_1 is numeric, then the corresponding estimated regression coefficient $\widehat{\beta}_1$ indicates by how much the estimated mean response $\widehat{\mathbb{E}}(y)$ changes for a unit increase in x_1, provided all the other predictors are held fixed. To see that, we can write $\widehat{\mathbb{E}}(y|x_1 + 1) - \widehat{\mathbb{E}}(y|x_1) = \widehat{\beta}_0 + \widehat{\beta}_1(x_1 + 1) + \widehat{\beta}_2 x_2 + \cdots + \widehat{\beta}_k x_k - (\widehat{\beta}_0 + \widehat{\beta}_1 x_1 + \widehat{\beta}_2 x_2 + \cdots + \widehat{\beta}_k x_k) = \widehat{\beta}_1$.

- If a predictor variable x_1 is an *indicator variable* (also termed *0-1 variable* or *dummy variable*), then $\widehat{\beta}_1$ has the meaning of the difference between the estimated mean response $\widehat{\mathbb{E}}(y)$ for $x_1 = 1$ and that for $x_1 = 0$, controlling for the other predictors. Indeed, $\widehat{\mathbb{E}}(y|x_1 = 1) - \widehat{\mathbb{E}}(y|x_1 = 0) = \widehat{\beta}_0 + \widehat{\beta}_1 \cdot 1 + \widehat{\beta}_2 x_2 + \cdots + \widehat{\beta}_k x_k - (\widehat{\beta}_0 + \widehat{\beta}_1 \cdot 0 + \widehat{\beta}_2 x_2 + \cdots + \widehat{\beta}_k x_k) = \widehat{\beta}_1$.

If the link function in (1.7) is not the identity, the interpretation of regression coefficients is analogous to the above, but is done in terms of $g(\widehat{\mathbb{E}}(y))$, the link function of the estimated mean response.

1.6 Model Goodness-of-Fit Check

According to the theory of generalized linear regression modeling, relative goodness-of-fit of several models may be compared based on a number of criteria, including the Akaike information criterion[3], corrected Akaike information criterion[4], and Schwarz Bayesian information criterion[5]. These three criteria are built upon the log-likelihood function $\ln L$ of the model since the larger its value, the better the fit of the model. However, it is known that a perfect fit may be achieved by introducing a large number of predictors into the model. Therefore, these criteria penalize for fitting too many predictors. Let p denote

---

[3] Akaike, H. (1974)."A new look at the statistical model identification". *IEEE Transactions on Automatic Control*, 19 (6): 716 – 723.

[4] Sugiura, N. (1978). "Further analysis of the data by Akaike's information criterion and the finite corrections". *Communications in Statistics – Theory and Methods*, A7: 13 – 26.

[5] Schwarz, G. (1978). "Estimating the dimension of a model". *Annals of Statistics*, 6(2): 461 – 464.

the number of unknown parameters in the regression model that have to be estimated from the data. For instance, in the case of the general linear regression (1.1), there are a total of $k + 2$ parameters (β_0 through β_k, and σ).

The *Akaike Information Criterion (AIC)* value is defined as

$$AIC = -2 \ln L + 2\,p. \tag{1.9}$$

The *Corrected Akaike Information Criterion (AICC)* value is given by the formula

$$AICC = -2 \ln L + 2p \frac{n}{n - p - 1}. \tag{1.10}$$

The AICC represents the AIC with a correction for a small sample size n.

The *Schwarz Bayesian Information Criterion (BIC)* (or, simply, *Bayesian Information Criterion*) value is computed as

$$BIC = -2 \ln L + p \ln(n). \tag{1.11}$$

For each of the three criteria, a model with the smallest value has the best fit. Note that these criteria don't provide goodness-of-fit measure in absolute terms, only relative to other models. When several regression models for the same response variable are developed that are based on different distributions and/or link functions, then the best fitted model may be chosen according to the above criteria.

In absolute terms, though, the goodness-of-fit of a given model is based on the *deviance test* (also known as the *asymptotic likelihood ratio test*).[6] In this test, the null hypothesis is that the *null model* has a better fit, and the alternative hypothesis is that the fitted model is better. Unless stated otherwise, the null model is the *intercept-only model*, which contains no predictors $x_1, \ldots, x_k$. The test statistic is called the *deviance*, and is defined as -2 multiplied by the difference in log-likelihoods of the null and fitted models, that is,

$$deviance = -2\Big(\ln L(null\ model) - \ln L(fitted\ model) \Big).$$

Under H_0, the test statistic has a chi-squared distribution with k degrees of freedom, and the p-value is calculated as the area under the density curve

---

[6]Introduced in Wilks, S. S. (1938)."The Large-sample distribution of the likelihood ratio for testing composite hypotheses". *The Annals of Mathematical Statistics*, 9(1): 60 – 62.

above the test statistic. The number of degrees of freedom is calculated as the difference between the number of parameters of the fitted and null models.

1.7   Predicted Response

For a general linear regression model, the formula for the fitted mean response (1.6) may be used to predict the value y^0 of the response variable for a fixed set of predictors $x_1^0, \ldots, x_k^0$. The prediction is carried out by calculating

$$y^0 = \widehat{\beta}_0 + \widehat{\beta}_1 x_1^0 + \cdots + \widehat{\beta}_k x_k^0. \tag{1.12}$$

For a generalized linear regression model, in view of (1.8), the *predicted response* is computed as

$$y^0 = g^{-1}\left(\widehat{\beta}_0 + \widehat{\beta}_1 x_1^0 + \cdots + \widehat{\beta}_k x_k^0\right). \tag{1.13}$$

1.8   SAS Implementation

Prior to fitting a general linear regression, it is wise to verify that the distribution of the response variable is indeed normal. To this end, we can plot a histogram for the measurements of y and overlay it with a normal probability density curve, which mean and standard deviation are estimated from the data. This can be done with the following syntax:

```
proc univariate data=data_name;
    var response_name;
    histogram /normal;
run;
```

The output contains the requested graph as well as test statistics and p-values for several normality tests, which are goodness-of-fit tests for normal distribution with the null hypothesis being that the distribution is normal and the alternative being that it is non-normal. SAS outputs the results of the three tests: Kolmogorov-Smirnov test, Cramer-von Mises test, and Anderson-Darling test. For these tests, p-values larger than 0.05 indicate normality of the distribution of the response variable at the 5% level of significance.

Once normality is established, the general linear regression model could be fitted to the data. We describe next how to do that.

A generalized linear regression model may be fitted using the procedure `genmod` with a specified distribution of the response variable and the type of link function. As a special case, the general linear regression model is fitted with the normal distribution of y and the identity link function. Both of these are, in fact, defaults in SAS. In what follows, however, we will always specify the distribution and link function as a reminder to ourselves.

To include computation of predicted response for fixed values of predictor variables, the incomplete data may be added as the last row of the data set with a dot in place of the missing response, and the following syntax can be run on this data set:

```
proc genmod data=data_name;
  class catpredictor1_name (ref='level_name') catpredictor2_name
        (ref='level_name') ...;
model response_name = <list of predictors>/dist=dist_name link=link_type;
      output out=outdata_name p=predicted_response_name;
run;

proc print data=outdata_name;
run;
```

• All categorical predictors *catpredictor1_name*, *catpredictor2_name*, ... must be listed in the `class` statement, and the reference level should be specified. If the appropriate reference level is not specified, then SAS chooses the last level in alphabetical order as a reference level for this variable. For ease of practical interpretation, it is recommended that the reference level for each categorical predictor be chosen so that the estimated regression coefficients for all the other levels of this variable have the same sign (typically, positive, sometimes negative, depending on what response variable is modeled).
• In the `model` statement, all predictors are listed, categorical or numeric, separated by a space.
• SAS automatically outputs the log-likelihood function of the fitted model, the values of the AIC, AICC, and BIC criteria, the estimated beta coefficients along with p-values for significance tests, and the estimated dispersion parameter.
• To conduct the goodness-of-fit test for a particular model, first fit the null model by running the code:

```
proc genmod data=data_name;
    model response_name = / dist=dist_name link=link_type;
run;
```

The deviance and the corresponding p-value may then be computed on a calculator, or in SAS. In SAS, for a given value of *deviance* and degrees of freedom *df*, the expression for the p-value becomes `1-probchi`(*deviance, df*).

• The predicted response for the specified values of x variables can be found in the last row of the printed data set *outdata_name*.

1.9 R Implementation

To plot a histogram for the response variable and to check for normality of its distribution, use the following syntax:

```
install.packages("rcompanion")
library(rcompanion)
```

```
plotNormalHistogram(data.name$response.name)
shapiro.test(data.name$response.name)
```

The bell-shaped curve will be plotted on the same graph with the histogram. In addition, the results of the Shapiro-Wilk test for normality of distribution will be displayed. As expected, for this test, the null hypothesis claims that the distribution is normal, while alternative asserts otherwise. A p-value larger than 0.05 will indicate normality at the 5% level of significance.

Further, the R function `glm()` can be used to fit a generalized linear regression model. The basic script is:

`summary`(*fitted.model.name*<- `glm`(*response.name* $\sim$ *x1.name* + $\cdots$ + *xk.name*, `data=`*data.name*, `family=`*dist.name*(`link=`*link.type*)))

• Actually, the above script will work only if all the listed predictors are numeric. Categorical predictors with numeric levels should be included as `as.factor`(*predictor.name*).

• For categorical predictors with character levels, the lowest alphabetically level is chosen as the reference. If you wish to change the reference category, apply the function `relevel()` first in the following manner:

releveled.predictor.name<- `relevel`(*data.name*$*predictor.name*, `ref="`*level.name*").

- The function `summary()` outputs regression coefficients, corresponding p-values for the tests of significance and the value for the AIC. The value of BIC can be requested by typing `BIC(`*fitted.model.name*`)`. The value of AICC is not computed automatically. It can be calculated as

$$\texttt{AICC<- -2*logLik(}fitted.model.name\texttt{)+2*}p\texttt{*}n\texttt{/(}n\texttt{-}p\texttt{-1)}$$

with the appropriate values of p and n. Here `logLik(`*fitted.model.name*`)` refers to the log-likelihood function of the fitted model.

- The estimate of sigma may be printed by specifying `sigma(`*fitted.model.name*`)`. The estimates of sigma produced by R and SAS are related as

$$\widehat{\sigma}_R = \widehat{\sigma}_{SAS} \sqrt{\frac{n}{n-k-1}}.$$

- To carry out the deviance test, first fit the null model and then compute the deviance and the p-value by adding the following lines to the script:

```
null.model.name<- glm(response.name ~ 1, data=data.name,
family=dist.name(link=link.type))
print(deviance.name<- -2*(logLik(null.model.name)
-logLik(fitted.model.name)))
print(p.value<- pchisq(deviance.name, df=value, lower.tail=FALSE))
```

- To use the fitted model for prediction, implement the function `predict()`. The values for categorical predictors are placed in double quotation marks, whereas for numeric predictors no quotation marks are needed. The syntax is:

```
predict(fitted.model.name, data.frame(catpredictor1.name="value", ...,
numpredictork.name=value)).
```

1.10   Example

EXAMPLE 1.1. A survey of 48 employees of a large company was conducted with the purpose of determining how satisfied they are with their jobs. Such demographic variables as gender, age, and education (Bachelor, Master, or Doctoral degree) were recorded. The total satisfaction score was calculated as a sum of scores on 20 questions on a 5-point Likert scale. We use these data to develop a regression model that relates the job satisfaction score to the other variables. First, we plot the histogram for the scores, and conduct the normality tests. We submit the following code:

```
data job_satisfaction;
  input gender$ age educ$ score @@;
cards;
M   53   doctoral   93    M   48   bachelor   66    M   47   master     82
M   34   bachelor   95    F   35   master     78    M   25   master     62
F   31   bachelor   87    F   25   master     76    M   26   master     71
M   58   bachelor   80    F   41   master     75    F   55   bachelor   75
M   40   bachelor   93    M   22   bachelor   96    F   49   master     63
F   39   bachelor   77    F   56   doctoral   73    F   49   master     56
M   45   master     77    M   23   bachelor   77    M   46   bachelor   79
M   25   master     94    F   62   bachelor   76    M   32   master     90
F   42   master     74    M   53   bachelor   92    F   29   master     91
F   47   bachelor   87    F   47   bachelor   55    M   27   bachelor   92
F   30   master     69    M   36   master     62    M   64   bachelor   77
F   40   bachelor   65    F   34   master     81    M   48   bachelor   64
M   46   bachelor   97    M   43   bachelor   80    F   37   bachelor   60
F   33   master     81    F   55   doctoral   68    M   22   bachelor   100
M   24   bachelor   68    M   54   bachelor   76    M   42   doctoral   81
F   63   bachelor   51    F   32   master     75    M   51   doctoral   81
;

proc univariate;
 var score;
  histogram/normal;
run;
```

The histogram and p-values for normality tests are given below.

Test	p Value
Kolmogorov-Smirnov	>0.150
Cramer-von Mises	>0.250
Anderson-Darling	>0.250

Judging by the histogram and the large p-values, the distribution of the scores is normal. Now we are ready to fit a general linear model. We write

```
proc genmod;
 class gender(ref='F') educ(ref='master');
  model score=gender age educ/dist=normal link=identity;
run;
```

The following estimators of the regression coefficients with the corresponding p-values for the test of significance are outputted. The value of the log-likelihood function and the estimate $\hat{\sigma}$ of the standard deviation (termed Scale) are also

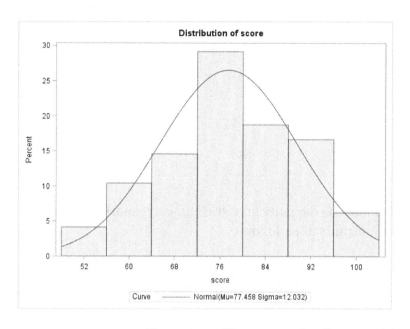

Figure 1.1: Histogram for Score in SAS

given.

Log Likelihood   -180.4720

Parameter	Estimate	Pr > ChiSq
Intercept	84.2229	<.0001
gender M	7.4876	0.0184
age	-0.3330	0.0216
educ bachelor	3.8754	0.2743
educ doctoral	7.5983	0.1938
Scale	10.3905	

From this output, the fitted regression model is $\widehat{\mathbb{E}}(score) = 84.2229 + 7.4876 * male - 0.3330 * age + 3.8754 * bachelor + 7.5983 * doctoral$, and the estimated standard deviation of the error is $\widehat{\sigma} = 10.3905$. Only gender and age, however, are statistically significant predictors of job satisfaction score at the 5% level of significance, since the corresponding p-values are less than 0.05. In this model, the estimated mean satisfaction score for men is 7.4876 points larger than that for women. In addition, with a one-year increase in age, the estimated average job satisfaction score is reduced by 0.333 points.

To address the question of whether this model fits the data well, we note from the output that the log-likelihood function for the fitted model is equal to

−180.4720. Next we fit the intercept-only model by submitting these statements:

```
proc genmod;
   model score=/dist=normal link=identity;
run;
```

Log Likelihood   -187.0063

Now we are ready to calculate the deviance statistic and carry out the goodness-of-fit test. The code and output are as follows:

```
data deviance_test;
 deviance=-2*(-187.0063-(-180.4720));
   pvalue=1-probchi(deviance,4);
run;

proc print;
run;
```

deviance   pvalue
 13.0686   0.010945

The p-value is less than 0.05, thus we conclude that at the 5% level of significance, the fitted model has a better fit.

Lastly, suppose we would like to use the fitted model to predict the job satisfaction score for a new female employee of this company who is 40 years of age and has a bachelor degree. To this end, we compute $y^0 = 84.2229 - 0.3330*40 + 3.8754 = 74.7783$. SAS outputs a very similar prediction, when the following statements are submitted. Here we first create a data set containing the values of the predictors, then turn it into the last row in the data set on which the model is run.

```
data prediction;
input gender$ age educ$;
cards;
F 40 bachelor
;

data job_satisfaction;
   set job_satisfaction prediction;
```

```
run;

proc genmod;
  class gender(ref='F') educ(ref='master');
   model score=gender age educ/dist=normal link=identity;
    output out=outdata p=pred_score;
run;

proc print data=outdata (firstobs=49 obs=49);
var pred_score;
run;
```

pred_score
   74.7802

The annotated R script that is needed to conduct all of the above analyses is given below along with the informative output.

```
job.satisfaction.data<- read.csv(file="./Example1.1Data.csv",
header=TRUE, sep=",")

#plotting histogram with fitted normal density
install.packages("rcompanion")
library(rcompanion)

plotNormalHistogram(job.satisfaction.data$score)

#testing normality of distribution
shapiro.test(job.satisfaction.data$score)
```

Shapiro-Wilk normality test
W = 0.97436, p-value = 0.3706

```
#specifying reference levels
educ.rel<- relevel(job.satisfaction.data$educ, ref="master")

#fitting general linear model
summary(fitted.model<- glm(score ~ gender + age + educ.rel,
data=job.satisfaction.data, family=gaussian(link=identity)))
```

Coefficients:

	Estimate	$Pr(>|t|)$
(Intercept)	84.2229	<2e-16

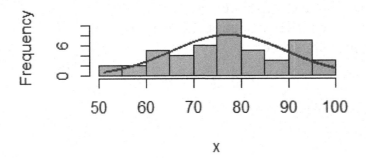

Figure 1.2: Histogram for Score in R

gender.relM	7.4876	0.0309
age	-0.3330	0.0352
educ.relbachelor	3.8754	0.3066
educ.reldoctoral	7.5983	0.2254

```
#outputting estimated sigma
sigma(fitted.model)
```

10.97801

```
#checking model fit
intercept.only.model<- glm(score ~ 1,
data=job.satisfaction.data,family=gaussian
(link=identity))
print(deviance<- -2*
(logLik(intercept.only.model)-logLik(fitted.model)))
```

13.06871

```
print(p.value<- pchisq(deviance, df=4, lower.tail=FALSE))
```

0.01094489

```
#using fitted model for prediction
print(predict(fitted.model, data.frame(gender="F", age=40,
educ.rel="bachelor")))
```

74.78019

□

Exercises for Chapter 1

EXERCISE 1.1. Show that the general linear regression model (1.1) is a special case of the generalized linear regression, that is, show that the normal distribution of the response variable y belongs to the exponential family of distributions (1.3) with $\theta = \mu$ and $\phi = \sigma^2$.

EXERCISE 1.2. A small-scale clinical trial is conducted to compare the efficacy of two drugs (A and B) in reduction of excess body weight. Drug (A or B), age, and gender were recorded at the baseline. The percent excess body weight loss (EWL) was recorded 3 months into the study. The data are available on 32 subjects:

Drug	Age	Gender	EWL	Drug	Age	Gender	EWL
A	49	F	14.2	A	44	F	6.7
A	54	M	25.4	B	52	F	29.4
A	37	F	14.1	B	51	M	21.9
A	43	F	20.0	B	44	F	23.6
A	57	M	11.7	B	53	F	23.8
A	48	M	16.6	B	55	M	7.4
A	34	F	15.9	B	30	F	23.1
A	51	F	17.4	B	47	M	16.8
A	54	F	22.8	B	26	M	14.1
A	45	F	16.7	B	56	F	24.6
A	36	M	12.7	B	28	F	17.8
A	57	M	15.0	B	34	M	27.8
A	44	M	8.4	B	43	M	10.6
A	56	M	11.2	B	55	M	26.8
A	44	M	17.3	B	52	F	15.7
A	47	M	20.5	B	54	F	23.7

(a) Verify normality of the response variable, then fit the linear regression model to the data. State the fitted model. Give estimates for all parameters.
(b) Which regression coefficients turn out to be significant at the 5%? Discuss goodness-of-fit of the model.
(c) Is one of the drugs more efficient for weight loss than the other? Interpret all estimated significant coefficients.
(d) According to the model, what is the predicted percent decrease in excess body weight for a 35-year old male who is taking drug A?

EXERCISE 1.3. A person is thinking of buying a new car. He conducts an on-line search and collects information on makes and models that he likes. He conjectures that the following car characteristics may potentially influence its price: body style (coupe, hatchback, or sedan), country of manufacture (USA, Germany, or Japan), highway mileage (in mpg), number of doors (2 or 4), and whether the interior is leather or not. The data for these variables and the price (in U.S. dollars) are given below for 27 cars.

Bodystyle	Country	Hwy	Doors	Leather	Price
coupe	USA	26	4	no	17,445
coupe	USA	40	4	no	23,500
coupe	USA	35	2	no	19,600
coupe	Germany	37	4	no	23,400
coupe	Germany	25	4	no	24,100
coupe	Germany	24	2	no	12,400
coupe	Japan	26	2	no	13,300
coupe	Japan	27	4	no	15,550
coupe	Japan	20	4	yes	29,345
hatchback	USA	30	2	no	12,540
hatchback	USA	39	4	no	17,595
hatchback	USA	38	2	no	17,300
hatchback	Germany	38	4	no	17,800
hatchback	Germany	32	4	no	22,500
hatchback	Germany	34	4	no	20,300
hatchback	Japan	38	4	yes	27,300
hatchback	Japan	38	2	yes	23,300
hatchback	Japan	38	2	yes	29,300
sedan	USA	29	4	no	32,000
sedan	USA	25	2	yes	34,200
sedan	USA	33	4	yes	33,395
sedan	Germany	40	4	no	22,850
sedan	Germany	23	2	yes	36,000
sedan	Germany	25	4	no	19,900
sedan	Japan	40	4	yes	36,700
sedan	Japan	35	4	yes	31,600
sedan	Japan	37	4	no	24,600

(a) Reduce the car price by the factor of 1000. Check that the distribution of the price is normal. Fit a general linear regression model to predict the price of a car. Write down the fitted model, specifying all estimated parameters.

(b) How good is the model fit? Discuss significance of the regression coefficients.

(c) Interpret the estimates of those regression coefficients that differ significantly from zero.

(d) What is the predicted price of a sedan made in USA that has 4 doors, leather seats, and runs 30 mpg on highway?

EXERCISE 1.4. Fifty people were surveyed randomly regarding the number of hours of quality sleep they normally get per night. Additional measurements on surveyed participants were age (in years), gender(M/F), number of minutes per day spent having personal quiet time, number of children under 5 years of age in the household, daily stress level (on a scale of 1 to 10), current job status(full/part/unemployed/student), number of physical activities per week, and number of months since last vacation or a weekend get-away. The data are as follows:

Age	Gender	Quiet time	N of children	Stress level	Job status	N of activities	Months since vacation	Sleep hours
62	F	60	1	5	unempl	1	15	7.7
28	F	15	1	6	unempl	5	11	5.3
50	M	15	0	5	unempl	1	19	6.4
36	M	60	1	6	full	1	21	7.7
56	F	50	0	3	part	4	5	7.6
26	F	80	0	7	student	9	8	8.3
48	M	180	0	5	full	0	6	6.4
55	M	40	0	8	full	8	23	7.0
44	M	180	1	3	part	6	20	9.6
49	F	5	0	7	unempl	5	15	5.5
29	M	60	2	5	student	5	7	7.7
56	M	10	1	4	unempl	4	17	5.7
46	F	40	1	7	part	3	3	7.4
41	F	5	2	6	full	9	10	6.2
22	M	15	0	8	full	4	3	6.3
36	F	45	2	5	part	8	14	7.5
54	F	120	1	8	part	7	10	8.5
42	F	60	3	1	full	9	11	6.3
58	F	5	1	7	full	1	17	5.3
33	M	100	2	1	full	9	5	8.3
50	F	2	2	6	full	3	12	5.1
59	M	30	2	5	full	2	6	6.9
32	M	30	1	8	full	5	9	6.9
50	M	60	2	8	part	8	13	8.0
56	F	10	0	3	unempl	7	7	6.1
42	F	240	0	1	part	8	21	8.8
58	F	10	2	7	full	9	4	6.2
57	F	15	1	6	full	2	16	6.3
30	F	30	0	2	full	8	9	8.3
54	M	20	2	8	full	6	7	6.5
57	M	45	2	4	full	7	18	7.5

(Continues on the next page)

(Continued from the previous page)

Age	Gender	Quiet time	N of children	Stress level	Job status	N of activities	Months since vacation	Sleep hours
45	F	120	0	9	part	2	13	6.6
33	F	40	1	6	unempl	9	24	7.0
56	F	120	0	5	part	2	20	8.7
59	F	60	2	9	part	4	19	8.1
41	M	60	2	3	student	2	3	7.5
62	M	40	0	1	unempl	0	2	8.6
29	M	15	1	7	unempl	3	20	6.3
34	F	30	0	7	unempl	9	0	6.6
32	F	20	3	7	unempl	2	8	7.8
46	F	20	2	3	unempl	9	18	7.9
45	M	60	0	2	unempl	0	22	9.0
23	M	45	0	6	part	4	12	7.6
38	M	60	4	5	full	3	5	7.8
45	M	30	0	5	unempl	9	7	6.8
63	F	40	0	6	unempl	5	5	7.3
27	F	120	0	4	student	1	16	7.3
30	F	45	0	7	part	8	10	7.7
34	F	5	3	6	full	0	4	6.0
62	M	10	0	10	part	8	11	6.0

(a) Show normality of the distribution of the number of hours of sleep per night. Regress the number of hours of sleep on all the given factors. Write explicitly what the fitted model is.

(b) How good is the model fit? What beta coefficients are significantly different from zero at the 5% level of significance?

(c) Interpret the estimated significant regression coefficients.

(d) Find the estimated number of hours of night's sleep that a 30-year old full-time mom of three children under the age of five has, if she gets 10 minutes a day for herself, walks to the park with her kids every day of the week, estimates her stress level as 7, and who hasn't gotten any vacation for one year.

EXERCISE 1.5. A 25-year-old student is training for a reverse triathlon (5-kilometer run, 13-mile bike, and 200-meter swim). He is interested in finding out what variables predict the total time spent on transitions. He obtains data for 420 people who participated in this triathlon the previous year, and chooses randomly 42 of them to do a regression analysis. The data presented below consist of age, gender, run time, time spent on transition 1 (T1), bike time,

time spent on transition 2 (T2), and swim time. All times are in minutes.

Age	Gender	Run	T1	Bike	T2	Swim
55	M	24.17	2.60	37.95	2.50	5.70
59	F	34.88	2.83	52.15	3.05	5.20
24	M	32.97	2.55	59.20	3.47	5.37
53	F	22.20	1.83	46.70	2.15	5.50
51	M	27.35	1.75	42.05	2.32	3.75
38	F	32.13	2.38	50.92	2.95	6.00
66	M	25.39	1.95	41.57	2.80	3.93
30	F	24.67	1.58	48.28	2.77	5.68
43	F	42.33	2.78	63.60	4.08	7.18
47	F	28.73	2.35	45.57	3.90	6.62
26	F	29.62	2.92	51.23	3.85	4.92
45	M	22.23	2.07	38.95	2.35	4.28
29	F	26.93	2.10	44.33	2.45	7.47
34	M	17.75	0.75	33.27	1.23	3.65
39	M	37.47	2.52	55.67	4.47	8.60
54	M	36.63	3.27	43.92	3.08	7.15
26	M	34.42	2.73	52.62	2.67	9.23
36	M	27.38	2.22	39.03	2.92	7.43
42	M	21.37	2.12	35.95	1.93	3.95
49	M	29.03	4.50	38.53	3.95	8.80
42	F	28.53	3.27	49.85	3.67	8.13
42	F	25.12	1.72	39.52	2.50	4.55
42	F	26.33	1.70	48.98	2.30	5.02
41	F	36.75	3.95	62.85	3.13	6.93
15	M	25.12	1.70	44.75	3.20	7.48
48	M	26.52	4.43	40.98	3.82	6.58
37	M	28.30	2.85	41.78	3.47	6.02
55	M	31.25	2.70	43.43	3.25	5.25
42	M	24.38	1.45	37.13	1.83	3.70
25	M	33.45	2.25	51.38	4.03	7.45
12	F	27.62	2.23	55.47	2.97	4.37
23	F	28.55	2.17	54.57	2.55	7.90
49	M	33.88	2.77	54.82	3.87	6.90
53	F	26.97	1.77	42.33	3.40	6.58
45	F	26.58	1.65	44.30	2.52	5.40
33	F	32.32	2.10	54.87	2.32	6.25
63	M	40.53	3.78	69.75	3.83	12.17
50	M	33.68	3.07	43.57	3.13	5.77
43	F	34.93	2.58	62.35	2.95	7.92
24	M	22.88	1.82	39.55	2.12	4.03
44	M	29.25	2.47	45.60	2.75	9.18
51	F	36.98	3.70	46.58	5.18	7.60

(a) Compute the total time spent on both transitions. Verify normality of the distribution of this variable, and fit a general linear regression model. Specify the fitted model.

(b) Discuss the model fit. Are all the predictors in that model significant at the 5% significance level?

(c) Interpret only the estimated significant regression coefficients of this model.

(d) What is the predicted total time at transitions for the student, if his best result in 5-kilometer run is 27:32, 13-mile bike is 56:17, and 200-meter swim is 8:46?

EXERCISE 1.6. A cardiologist conducts a study to find out what factors are good predictors of elevated heart rate (HR) in her patients. She measures heart rate at rest in 30 patients on their next visit, and obtains from the medical charts additional information on their age, gender, ethnicity(Black, Hispanic, or White), body mass index (BMI), and the number of currently taken heart medications. She also obtains the air quality index (AQI) for area of residence of her patients (unhealthy/moderate/good). The data follow.

Age	Gender	Ethnicity	BMI	Meds	AQI	HR
48	F	Black	29.9	0	good	76
56	F	White	22.9	3	unhealthy	112
67	F	White	23.4	1	good	94
82	M	Black	29.7	0	good	92
64	F	White	31.4	3	good	97
58	M	White	18.9	2	moderate	79
72	F	Black	25.2	0	moderate	114
70	F	Black	25.9	1	moderate	115
54	M	Hispanic	29.6	0	moderate	80
57	F	Hispanic	20.2	2	good	81
50	F	Black	23.9	1	unhealthy	97
59	F	Hispanic	22.6	0	good	86
61	M	Hispanic	32.8	1	good	84
69	M	Hispanic	24.1	2	unhealthy	94
65	F	Black	23.4	2	moderate	114
66	F	Hispanic	27.8	3	good	82
74	M	White	32.4	1	moderate	97
66	M	Hispanic	22.9	2	good	86
53	M	Hispanic	25.2	0	good	84
55	M	Hispanic	24.6	0	moderate	94
73	F	Hispanic	24.8	3	moderate	105
45	F	Hispanic	19.0	2	unhealthy	83
71	F	White	20.3	2	unhealthy	111
63	M	Black	23.8	2	unhealthy	108
71	F	White	21.5	2	moderate	100
62	M	Hispanic	27.4	3	good	79
44	F	Hispanic	17.2	0	unhealthy	86
49	M	White	17.1	1	good	75
63	M	Black	28.0	2	good	91
65	F	Hispanic	22.2	1	moderate	106

(a) Check that the measurements for the heart rate are coming from a normal distribution. Fit the regression model and specify all estimated parameters.

(b) Discuss the goodness-of-fit of the model. What variables are significant predictors of heart rate at the 5% level of significance?

(c) Give interpretation of the estimated statistically significant regression coefficients.

(d) Compute the predicted heart rate of a 50-year-old Hispanic male who has a BMI of 20, is not taking any heart medications, and resides in an area with a moderate air quality.

Chapter 2

Regression Models for Response with Right-skewed Distribution

Suppose the response variable y is continuous, assumes only positive values, and its histogram doesn't appear to be roughly symmetrical and bell-shaped (that is, normal) but rather has a long right tail (skewed to the right). In this chapter, we talk about two possible approaches to modeling such a response: the Box-Cox transformation and gamma regression.

2.1 Box-Cox Power Transformation

2.1.1 Model Definition

If the density of the response variable y is right-skewed, a transformation may be applied to y to make its density look more normally shaped. This transformation is referred to as *Box-Cox power transformation*[1] (or, simply, *Box-Cox transformation*). The transformed response variable, denoted here by $\tilde{y}$, has the form:

$$\tilde{y} = \begin{cases} \dfrac{y^\lambda - 1}{\lambda}, & \text{if } \lambda \neq 0, \\ \ln y, & \text{if } \lambda = 0. \end{cases} \tag{2.1}$$

This is a well-defined transformation since values of y are assumed positive, which can always be achieved by adding an appropriate constant to all values of y. Note also that the way this transformation is defined makes it continuous in λ. Indeed, by the l'Hôpital's rule, $\displaystyle\lim_{\lambda \to 0} \frac{y^\lambda - 1}{\lambda} = \lim_{\lambda \to 0} y^\lambda \ln y = \ln y$.

---

[1]Introduced in Box, G. E. P. and Cox, D. R. (1964). "An analysis of transformations". *Journal of the Royal Statistical Society, Series B*, 26(2), 211 – 252.

The optimal value of λ is found by means of the maximum likelihood estimation. For a set of discrete values of λ, a linear model is fitted, where the transformed response $\tilde{y}$ is regressed on predictor variables $x_1, \ldots, x_k$. The value of λ that corresponds to the maximum of the likelihood function is chosen as the optimal value. However, the described optimization is carried out under the assumption that $\tilde{y}$ is normally distributed, which doesn't hold exactly, so, in practice, researchers "round off" the values of λ to result in several meaningful transformations. These recommended transformations are summarized in the table below.

Range for optimal λ	Recommended value of λ	Transformed $\tilde{y}$	Transformation name
$[-2.5, -1.5)$	-2	$\frac{1}{2}\left(1 - \frac{1}{y^2}\right)$	*inverse square*
$[-1.5, -0.75)$	-1	$1 - \frac{1}{y}$	*inverse (or reciprocal)*
$[-0.75, -0.25)$	-0.5	$2\left(1 - \frac{1}{\sqrt{y}}\right)$	*inverse square root*
$[-0.25, 0.25)$	0	$\ln y$	*natural logarithm*
$[0.25, 0.75)$	0.5	$2(\sqrt{y} - 1)$	*square root*
$[0.75, 1.5)$	1	$y - 1$	*linear*
$[1.5, 2.5]$	2	$\frac{1}{2}(y^2 - 1)$	*square*

2.1.2 Fitted Model

Let λ denote the recommended value from the table above. The fitted mean for the Box-Cox transformed response is

$$\widehat{\mathbb{E}}(\tilde{y}) = \widehat{\mathbb{E}}\left(\frac{y^\lambda - 1}{\lambda}\right) = \widehat{\beta}_0 + \widehat{\beta}_1 x_1 + \cdots + \widehat{\beta}_k x_k. \tag{2.2}$$

2.1.3 Interpretation of Estimated Regression Coefficients

By (2.2), for a continuous predictor x_1, the estimated regression coefficient $\widehat{\beta}_1$ represents the change in estimated mean of the transformed response $\widehat{\mathbb{E}}(\tilde{y})$ when x_1 is increased by one unit, given that the other predictors stay fixed. If x_1 is an indicator variable, then $\widehat{\beta}_1$ is interpreted as the difference between the estimated mean of the transformed response $\widehat{\mathbb{E}}(\tilde{y})$ for $x_1 = 1$ and that for $x_1 = 0$, when the other predictors are kept unchanged.

2.1.4 Predicted Response

By (2.2), the predicted response y^0 for some given values of predictors $x_1^0, \ldots, x_k^0$ is

$$y^0 = \left(\lambda(\widehat{\beta}_0 + \widehat{\beta}_1 x_1^0 + \cdots + \widehat{\beta}_k x_k^0) + 1 \right)^{1/\lambda}.$$

2.1.5 SAS Implementation

The Box-Cox transformation may be performed via procedure `transreg` with the following syntax:

```
proc transreg;
    model BoxCox(response_name) = identity(<list of predictors>);
run;
```

- `BoxCox(·)` specifies the Box-Cox transformation.
- `identity(·)` specifies the identity transformation, since we assume no transformations are applied to the predictor variables. Within this function, predictors should be listed separated by spaces.
- Since the `class` statement is not allowed in this procedure, all categorical predictors must be 0-1 variables. Suppose we have a variable *variable_name* with levels 'A', 'B', and 'C', and we want to include it into the model with the level 'C' as the reference level. We proceed to creating two indicator variables by running the two lines of code in the `data` statement:

levelA_name=(*variable_name*='A');
levelB_name=(*variable_name*='B');

The general rule is that if a categorical predictor has c levels, then one of them must be chosen as a reference level, and $c - 1$ indicator variables must be created and entered into the model.

- SAS outputs a column of lambda values ranging between -3 and 3 with a step of 0.25, and a column of the corresponding values of the log-likelihood function for the fitted model. The optimal value of λ that corresponds to the maximum log-likelihood can be found by visual inspection. Once the optimal λ is known, Table 2.1 above can be used to pick the recommended transformation.

Further, in SAS, the transformed response variable may be computed and the general linear regression may be run on that variable. Prior to running the model, however, it is advisable to plot a histogram for the transformed variable, and to conduct the formal testing for normality of its distribution.

2.1.6   R Implementation

In R, the library MASS (stands for "Modern Applied Statistics with S") needs to be called to find the optimal λ for a Box-Cox transformation. First the Box-Cox transformation is applied, and a column of lambda values is created along with a column of corresponding values of the profile log-likelihood function (which is proportional to the full log-likelihood function up to an additive constant). To match SAS output, we can request that the lambda values range between -3 and 3 with a step of 1/4, and that no interpolation is applied. Note that all categorical predictors in the function boxcox() should be re-leveled first, using the function relevel(). The syntax for the Box-Cox transformation is:

```
install.packages("MASS")
library(MASS)
```

BoxCox.fit.name <- boxcox(*response.name* ~ *x1.name* + ...
+ *xk.name*, data=*data.name*, lambda=seq(-3,3,1/4), interp=FALSE)

In order to extract the value of lambda that corresponds to the largest value of the profile log-likelihood function, first one has to create a data frame with lambda and profile log-likelihood function columns as follows.

BoxCox.data.name <- data.frame(*BoxCox.fit.name*$x, *BoxCox.fit.name*$y)

Next, the two columns should be sorted in descending order with respect to the profile log-likelihood function values, so that the optimal lambda appears in the top row. The script that does that is below.

ordered.data.name <- *BoxCox.data.name*[with(*BoxCox.data.name*, order(-*BoxCox.fit.name*.y)),]

The final step is to display the value of the optimal lambda in the top row.

ordered.data.name[1,]

Once the value of lambda is identified, Table 2.1 should be used to determine the recommended value of λ. Then the transformation (2.1) should be carried out, its normality checked, and the general linear regression model fitted to the transformed response.

2.1.7 Example

EXAMPLE 2.1. A real estate specialist is interested in modeling house prices in a certain U.S. region. He suspects that house prices depend on such characteristics as the number of bedrooms, number of bathrooms, square footage of the house, type of heating (central/electrical/none), presence of air conditioner (A/C) (yes/no), and lot size. He obtains the data on 30 houses currently on the market.

The SAS code below conducts the appropriate analysis. To avoid working with large numbers, we divide price by 10,000, and divide square footage and lot size by 1000. Also, since the variables that describe the type of heating and presence or absence of air conditioner are categorical, we create appropriate indicator variables. The levels 'none' for heating and 'no' for A/C are chosen as references. The data statement in SAS looks like this:

```
data real_estate;
input price beds baths sqft heating$ AC$ lot;
price10K=price/10000;
sqftK=sqft/1000;
central=(heating='central');
electric=(heating='electric');
ACyes=(AC='yes');
lotK=lot/1000;
cards;
669000   3   2     1733   central    no    5641
715000   4   3.5   1812   none       yes   4995
634900   5   3     2217   none       no    8019
640000   3   2     1336   none       no    7283
966000   5   3     4000   central    no    7424
889000   3   2     2005   central    no    7130
745000   4   3.5   2276   none       no    7936
685000   2   1.5   1018   central    yes   6141
549500   2   1     920    central    no    5545
868999   5   2.5   1670   electric   yes   5750
624900   3   2     1519   electric   no    8267
549900   2   1     956    none       no    4978
589900   3   2     1601   central    no    5005
```

```
829000    5  3     2652  central   yes  5601
599900    4  2     1802  none      yes  5262
875000    6  2.5   3414  electric  yes  6534
635000    3  2     1565  central   no   5619
599999    2  1     832   none      no   5601
734997    3  2.5   1780  central   yes  5400
699999    3  2     1969  electric  no   5488
759000    4  2     1530  central   yes  6446
684900    3  2     1519  central   no   8267
888000    5  2.75  2039  central   yes  5976
599999    4  2     1513  electric  no   5937
565000    2  2     1616  central   no   5227
825000    3  2.5   1421  central   yes  5871
659900    3  2     1547  electric  yes  4791
746000    3  2     1130  central   no   5301
1089000   5  2.5   3314  central   yes  7129
1195499   5  3.5   3760  central   yes  6000
;
```

Prior to fitting a linear model, we would want to plot a histogram of the rescaled price to assess its deviation from normality. We run the code below to obtain the histogram with an overlaid normal density curve.

```
proc univariate;
  var price10K;
    histogram /normal;
run;
```

Goodness-of-Fit Tests for Normal Distribution

Test	p Value
Kolmogorov-Smirnov	>0.105
Cramer-von Mises	>0.024
Anderson-Darling	>0.018

The histogram exhibits a long right tail, suggesting that the distribution is right-skewed. Also, two out of the three tests refute normality of the response, since their p-values are below 0.05.

Our next step would be to transform the response variable by means of the Box-Cox transformation. We run procedure **transreg** as follows:

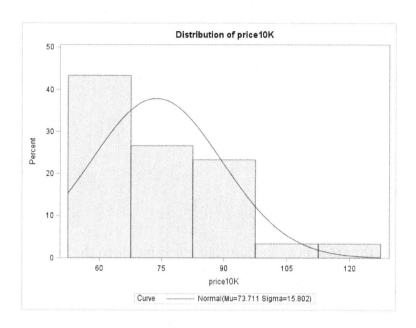

Figure 2.1: Histogram for Price10K in SAS

```
proc transreg data=real_estate;
model BoxCox(price10K) = identity(beds baths sqftK central
electric ACyes lotK);
run;
```

SAS outputs a column with λ values and a column with corresponding log-likelihood function values.

Lambda	Log Like
-3.00	-63.2544
-2.75	-62.7175
-2.50	-62.2403
-2.25	-61.8273
-2.00	-61.4832
-1.75	-61.2132
-1.50	-61.0229
-1.25	-60.9186
-1.00	-60.9068
-0.75	-60.9944
-0.50	-61.1884
-0.25	-61.4959
0.00	-61.9238
0.25	-62.4784
0.50	-63.1654

0.75	-63.9893
1.00	-64.9534
1.25	-66.0594
1.50	-67.3075
1.75	-68.6960
2.00	-70.2216
2.25	-71.8798
2.50	-73.6644
2.75	-75.5686
3.00	-77.5847

As seen from this output, the optimal value of λ that corresponds to the largest value of the log-likelihood function is -1, which, according to Table 2.1, is a convenient value of λ. Thus, an inverse transformation is recommended.

Further, we perform the recommended transformation and plot the histogram of the transformed response. The code that accomplishes this is as follows:

```
data real_estate;
 set real_estate;
  tr_price10K=1-(1/price10K);
run;

proc univariate;
 var tr_price10K;
  histogram /normal;
  run;
```

As seen on the histogram and confirmed by the large p-values (larger than 0.05) for the normality tests, the transformed response is indeed normally distributed.

Goodness-of-Fit Tests for Normal Distribution

Test	p Value
Kolmogorov-Smirnov	>0.150
Cramer-von Mises	>0.250
Anderson-Darling	>0.250

Finally, we fit the general linear model to the transformed response, regressing it on all the predictors. Here we use the original predictors and the `class` statement rather than the calculated indicator variables `central`, `electric`, and `ACyes`, which would give us an alternative solution.

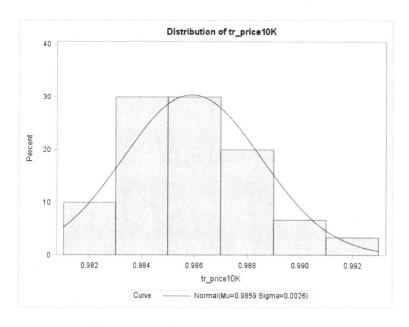

Figure 2.2: Histogram for Transformed Price10K in SAS

```
proc genmod;
class heating(ref='none') AC(ref='no');
model tr_price10K=beds baths sqftK heating
AC lotK/dist=normal link=identity;
run;
```

The relevant output is

Full Log Likelihood   157.9841

Analysis Of Maximum Likelihood Parameter Estimates

Parameter		Estimate	Pr > ChiSq
Intercept		0.9768	<.0001
beds		0.0002	0.7175
baths		0.0007	0.2512
sqftK		0.0012	0.0269
heating	central	0.0017	0.0052
heating	electric	0.0009	0.2425
heating	none	0.0000	.
AC	yes	0.0018	0.0034
AC	no	0.0000	.
lotK		0.0005	0.0704
Scale		0.0012	

Addressing the issue of goodness-of-fit of the fitted model, we can find in the output that log-likelihood for the fitted model is 157.9841. We obtain the log-likelihood function for the intercept-only model and conduct the deviance test. The code and appropriate output are:

```
proc genmod data=real_estate;
model tr_price10K=/dist=normal link=identity;
run;
```

Full Log Likelihood   136.2538

```
data deviance_test;
 deviance=-2*(136.2538-(157.9841));
   pvalue=1-probchi(deviance,7);
run;

proc print;
run;
```

deviance   pvalue
 43.4606   0.000000272

The p-value is way below 0.05. This, the fitted model has a good fit.

In the fitted model, only the square footage of a house, presence of central heater, and presence of air conditioner turn out to be significant predictors of the house price at the 5% level. If the square footage of the house were one thousand square feet larger, the estimated average inverse-transformed price (in \$10,000) would be 0.0012 units larger. For houses with central heater, the estimated inverse-transformed price is, on average, 0.0017 units larger than that for house with no heater. The estimated mean inverse-transformed price for a house with an air conditioner is 0.0018 units larger than that for a house with no air conditioner.

Next, we might want to use the model to predict the price of a house that has four bedrooms, two bathrooms, area of 1680 square feet, central heater, no A/C, and lot size of 5000 square feet. The formula that should be used for prediction is (show it!):

$$price^0 = 10000 * \left[1 - (0.9768 + 0.0002 * beds^0 + 0.0007 * baths^0 + 0.0012 * \frac{sqft^0}{1000}\right.$$

$$\left. + 0.0017 * central + 0.0009 * electric + 0.0018 * ACyes + 0.0005 * \frac{lot^0}{1000})\right]^{-1}.$$

Plugging the appropriate values into the formula above, we get that the predicted price is equal to

$$price^0 = 10000 * \left[1 - (0.9768 + 0.0002 * 4 + 0.0007 * 2 + 0.0012 * \frac{1680}{1000}\right.$$

$$\left.+0.0017 + 0.0005 * \frac{5000}{1000})\right]^{-1} = \$676, 406.93.$$

The prediction can also be requested in SAS via the following statements:

```
data prediction;
input beds baths sqftK heating$ AC$ lotK;
cards;
4 2 1.68 central no 5
;

data real_estate;
 set real_estate prediction;
 run;

proc genmod;
class heating(ref='none') AC(ref='no');
model tr_price10K=beds baths sqftK heating
AC lotK/dist=normal link=identity;
output out=outdata p=predicted;
run;

data outdata;
 set outdata;
 pred_price=10000/(1-predicted);
run;

proc print data=outdata (firstobs=31 obs=31);
var pred_price;
run;
```

The output is

```
pred_price
 658569.23
```

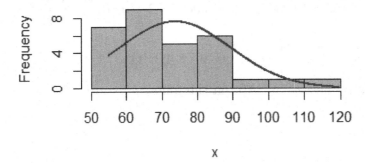

Figure 2.3: Histogram for Price10K in R

Thus, the predicted price of the house in SAS is \$658,569.23. Note that the discrepancy between the predicted prices computed by hand and in SAS is due to the round-off error. For calculations, SAS uses more digits in the estimates of the regression coefficients than it displays.

The R script that does the same tasks as described in this example is given below. Recall that in R, the Shapiro-Wilk test is used to verify normality. Relevant outputs are also presented.

```
real.estate.data<- read.csv(file="./Example2.1Data.csv",
header=TRUE, sep=",")


#rescaling variables and specifying reference categories
price10K<- real.estate.data$price/10000
sqftK<- real.estate.data$sqft/1000
heating.rel<- relevel(real.estate.data$heating, ref="none")
AC.rel<- relevel(real.estate.data$AC, ref="no")
lotK<- real.estate.data$lot/1000


#plotting histogram with fitted normal density
install.packages("rcompanion")
library(rcompanion)


plotNormalHistogram(price10K)


#testing for normality of distribution
shapiro.test(price10K)
```

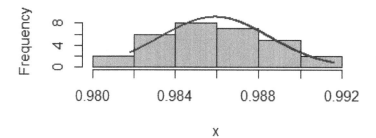

Figure 2.4: Histogram for Transformed Price10K in R

Shapiro-Wilk normality test
W = 0.89581, p-value = 0.006642

```
#finding optimal lambda for Box-Cox tranformation
install.packages("MASS")
library(MASS)

BoxCox.fit<- boxcox(price10K ~ beds + baths
+ sqftK + heating.rel + AC.rel + lotK, data=real.estate.data,
lambda = seq(-3,3,1/4), interp=FALSE)
BoxCox.data<- data.frame(BoxCox.fit$x, BoxCox.fit$y)
ordered.data<- BoxCox.data[with(BoxCox.data,
order(-BoxCox.fit.y)),]
ordered.data[1,]

#applying Box-Cox tranformation with lambda=-1
tr.price10K<- 1-(1/price10K)

#plotting histogram for tranformed response
plotNormalHistogram(tr.price10K)

#testint or normality of distribution
shapiro.test(tr.price10K)
```

Shapiro-Wilk normality test
W = 0.96903, p-value = 0.5131

```
#fitting general linear model to transformed response
summary(fitted.model<- glm(tr.price10K ~ beds + baths
+ sqftK + heating.rel + AC.rel + lotK,
data=real.estate.data, family=gaussian(link=identity)))
```

Coefficients:

	Estimate	Pr(>|t|)
(Intercept)	0.9767546	<2e-16
beds	0.0001599	0.7596
baths	0.0006527	0.3365
sqftK	0.0011973	0.0712
heating.relcentral	0.0017088	0.0257
heating.relelectric	0.0008685	0.3278
AC.relyes	0.0017836	0.0200
lotK	0.0004792	0.1355

```
#outputting estimated sigma
sigma(fitted.model)
```

0.001459009

```
#checking model fit
intercept.only.model<- glm(tr.price10K ~ 1,
family=gaussian(link=identity))
print(deviance<- -2*(logLik(intercept.only.model)
-logLik(fitted.model)))
```

43.46061

```
print(p.value<- pchisq(deviance, df=7, lower.tail=FALSE))
```

2.717569e-07

```
#using fitted model for prediction
pred.tr.price10K<-predict(fitted.model,
data.frame(beds=4, baths=2,
sqftK=1.68, heating.rel="central", AC.rel="no", lotK=5))
print(pred.price<- 10000/(1-pred.tr.price10K))
```

658569.2

Note that R outputs more digits in the estimates for beta coefficients. If we use these estimates when computing the predicted value, we would get a more accurate result. Indeed,

$$price^0 = 10000 * \left[1 - (0.9767546 + 0.0001599 * 4 + 0.0006527 * 2 + 0.0011973 * 1.68\right.$$
$$\left. + 0.0017088 + 0.0004792 * 5)\right]^{-1} = \$658,582.09.$$

□

2.2 Gamma Regression Model

2.2.1 Model Definition

A *gamma regression*[2] may be fitted to a positive response y with a right-skewed distribution. In this model, y has a gamma distribution with the density

$$f_Y(y) = \frac{y^{\alpha-1}}{\Gamma(\alpha)\,\beta^\alpha}\,e^{-y/\beta}, \quad \alpha,\,\beta > 0, \quad y > 0. \tag{2.3}$$

The expected value of y is $\mathbb{E}(y) = \alpha\,\beta$. The relation between the mean response and the predictor variables is modeled via a log link function:

$$\ln \mathbb{E}(y) = \beta_0 + \beta_1\, x_1 + \cdots + \beta_k\, x_k. \tag{2.4}$$

Here the parameters $\beta_0, \ldots, \beta_k$, and α are unknown and are estimated from the data by the method of maximum likelihood.

It can be shown (see Exercise 2.5) that the gamma distribution belongs to the exponential family of distributions since its density can be written in the form (1.3) with the location parameter $\theta = -1/(\alpha\beta)$ and the dispersion parameter $\phi = 1/\alpha$. Hence, the gamma regression is a generalized linear regression model with the log link function.

2.2.2 Fitted Model

The fitted mean response has the form:

$$\widehat{\mathbb{E}}(y) = \widehat{\alpha}\,\widehat{\beta} = \exp\{\widehat{\beta}_0 + \widehat{\beta}_1\, x_1 + \cdots + \widehat{\beta}_k\, x_k\}, \tag{2.5}$$

and the estimated dispersion parameter $\widehat{\phi} = 1/\widehat{\alpha}$. Thus, in the fitted model, the estimated parameters satisfy:

$$\widehat{\alpha} = 1/\widehat{\phi}, \quad \text{and} \quad \widehat{\beta} = \widehat{\phi}\,\exp\{\widehat{\beta}_0 + \widehat{\beta}_1\, x_1 + \cdots + \widehat{\beta}_k\, x_k\}.$$

2.2.3 Interpretation of Estimated Regression Coefficients

According to (2.5), estimated regression coefficients in a gamma regression yield the following interpretation:

---

[2]Discussed for the first time in Nelder, J.A. and Wedderburn, R.W.M. (1972). "Generalized linear models". *Journal of the Royal Statistical Society, Series A*, 135 (3): 370 – 384.

- If a predictor variable x_1 is numeric, then the corresponding estimated beta coefficient $\widehat{\beta}_1$ indicates by how much the natural logarithm of the estimated mean response $\ln \widehat{\mathbb{E}}(y)$ changes for a unit increase in x_1, provided all the other predictors stay unchanged. Or, equivalently, $(\exp\{\widehat{\beta}_1\} - 1) \cdot 100\%$ represents the percent change in estimated mean response for a unit increase in x_1. To obtain that, we write

$$\left(\widehat{\mathbb{E}}(y|x_1 + 1) - \widehat{\mathbb{E}}(y|x_1)\right) / \widehat{\mathbb{E}}(y|x_1) = \left(\exp\{\widehat{\beta}_0 + \widehat{\beta}_1(x_1 + 1) + \widehat{\beta}_2 x_2 + \cdots + \widehat{\beta}_k x_k\} \right.$$

$$\left. - \exp\{\widehat{\beta}_0 + \widehat{\beta}_1 x_1 + \cdots + \widehat{\beta}_k x_k\} \right) / \exp\{\widehat{\beta}_0 + \widehat{\beta}_1 x_1 + \cdots + \widehat{\beta}_k x_k\} = \exp\{\widehat{\beta}_1\} - 1.$$

- If a predictor variable x_1 is an indicator variable, then $\exp\{\widehat{\beta}_1\} \cdot 100\%$ has the meaning of the percent ratio of estimated mean response $\widehat{\mathbb{E}}(y)$ for $x_1 = 1$ and that for $x_1 = 0$, keeping the other predictors intact. Indeed,

$$\widehat{\mathbb{E}}(y|x_1 = 1) / \widehat{\mathbb{E}}(y|x_1 = 0) = \exp\{\widehat{\beta}_0 + \widehat{\beta}_1 \cdot 1 + \widehat{\beta}_2 x_2 + \cdots + \widehat{\beta}_k x_k\}/$$

$$\exp\{\widehat{\beta}_0 + \widehat{\beta}_1 \cdot 0 + \widehat{\beta}_2 x_2 + \cdots + \widehat{\beta}_k x_k\} = \exp\{\widehat{\beta}_1\}.$$

2.2.4   Predicted Response

From (2.5), the predicted response y^0 for a concrete set of predictors $x_1^0, \ldots, x_k^0$ can be found as $y^0 = \exp\{\widehat{\beta}_0 + \widehat{\beta}_1 x_1^0 + \cdots + \widehat{\beta}_k x_k^0\}$.

2.2.5   SAS Implementation

A gamma distribution may be fitted using `proc genmod` with `dist=gamma` and `link=log`.

2.2.6   R Implementation

The R function `glm()` will fit the gamma regression if `family=Gamma(link=log)` is specified. To obtain the predicted response, the option `type="response"` must appear in the function `predict()`.

2.2.7   Example

EXAMPLE 2.2. Consider the data `real_estate` in Example 2.1. We fit the gamma model and use it for prediction by running these lines of code:

```
proc genmod;
class heating(ref='none') AC(ref='no');
model price10K = beds baths sqftK heating AC lotK
    /dist=gamma link=log;
run;
```

The informative output is:

Full Log Likelihood   -100.0395

Analysis Of Maximum Likelihood Parameter Estimates

		Estimate	Pr > ChiSq
Intercept		3.6462	<.0001
beds		0.0091	0.7837
baths		0.0295	0.4887
sqftK		0.1165	0.0035
heating	central	0.1206	0.0084
heating	electric	0.0484	0.3879
heating	none	0.0000	.
AC	yes	0.1292	0.0052
AC	no	0.0000	.
lotK		0.0303	0.1326
Scale		113.4926	

Next, we check whether this model has a good fit. The code and relevant output are:

```
proc genmod;
 model price10K=/dist=gamma link=log;
run;
```

Full Log Likelihood -122.8174

```
data deviance_test;
 deviance=-2*(-122.8174-(-100.0395));
   pvalue=1-probchi(deviance,7);
run;

proc print;
run;
```

```
deviance        pvalue
45.5558   0.000000107
```

Since the p-value for the deviance test is very small, we conclude that the model with all the predictors fits the data well. The fitted model is

$$\widehat{\mathbb{E}}(price) = \exp\{3.6462 + 0.0091 * beds + 0.0295 * baths + 0.1165 * sqftK$$

$$+0.1206 * central + 0.0484 * electric + 0.0.1292 * ACyes + 0.0303 * lotK\},$$

and $\widehat{\alpha} = 1/113.4926 = 0.0088$. We see that the square footage of the house, central heater, and presence of A/C are significant predictors (the same as in the model in Example 2.1). If the square footage of a house were larger by one thousand square feet, then the estimated average price would be larger by $(\exp\{0.1165\} - 1) \cdot 100\% = 12.36\%$. For a house with central heater, the estimated mean of the price is $\exp\{0.1206\} \cdot 100\% = 112.82\%$ of that for a house with no heater. Houses with air conditioner have estimated average price $\exp\{0.1292\} \cdot 100\% = 113.79\%$ of that for houses where air conditioner is not installed.

Further, to predict the price of a house with four bedrooms, two bathrooms, area of 1680 square feet, central heating, no A/C, and lot size of 5000 square feet, we compute:

$$price^0 = 1000*\exp\{3.6462+0.0091*4+0.0295*2+0.1165*1.68+0.1206+0.0303*5\}$$

$$= \$673,174.84.$$

To reveal the predicted value in SAS, we run these statements:

```
data prediction;
input beds baths sqftK heating$ AC$ lotK;
cards;
4 2 1.68 central no 5
;

data real_estate;
  set real_estate prediction;
  run;

proc genmod;
class heating(ref='none') AC(ref='no');
model price10K = beds baths sqftK heating
```

```
     AC lotK/dist=gamma link=log;
output out=outdata p=pprice;
run;

data outdata;
 set outdata;
  pred_price=10000*pprice;
run;

proc print data=outdata (firstobs=31 obs=31);
var pred_price;
run;
```

The output is:

```
pred_price
 673237.87
```

The difference between the predicted values computed by hand and in SAS is
due to the fact that SAS uses more digits in the estimated beta coefficients
than are displayed in the output.

Finally, the R script below repeats the same steps we did in SAS. The outputted estimates are close to those produced by SAS.

```
real.estate.data<-read.csv(file="./Example2.1Data.csv",
header=TRUE, sep=",")

#rescaling variables and specifying reference categories
price10K<- real.estate.data$price/10000
sqftK<-real.estate.data$sqft/1000
heating.rel<- relevel(real.estate.data$heating, ref="none")
AC.rel<-relevel(real.estate.data$AC, ref="no")
lotK<-real.estate.data$lot/1000

#fitting gamma regression
summary(fitted.model<- glm(price10K ~ beds + baths + sqftK
+ heating.rel + AC.rel + lotK, data=real.estate.data,
family=Gamma(link=log)))
```

Coefficients:
                    Estimate   Pr(>|t|)

(Intercept)	3.646245	<2e-16
beds	0.009136	0.8183
baths	0.029540	0.5650
sqftK	0.116481	0.0241
heating.relcentral	0.120590	0.0371
heating.relelectric	0.048372	0.4717
AC.relyes	0.129186	0.0261
lotK	0.030274	0.2117

Dispersion parameter for Gamma family taken to be 0.01233554

```
#checking model fit
intercept.only.model<- glm(price10K ~ 1, family=Gamma(link=log))
print(deviance<- -2*(logLik(intercept.only.model)
-logLik(fitted.model)))
```

45.55644

```
print(p.value<- pchisq(deviance, df=7, lower.tail=FALSE))
```

1.066286e-07

```
#using fitted model for prediction
print(10000*predict(fitted.model, type="response",
data.frame(beds=4, baths=2, sqftK=1.68, heating.rel="central",
AC.rel="no", lotK=5)))
```

673237.9

□

Exercises for Chapter 2

EXERCISE 2.1. An intervention-control study on childhood obesity was conducted at a children's clinic. A cohort of 36 obese children, ages 6 through 16, were followed for 9 months. The intervention consisted of educational sessions for parents and vigorous exercise activities for kids. The control group participants were provided with resources regarding other active and healthy lifestyle programs offered in their community. Their gender(M/F), age (in years), group (intervention Tx or control Cx), and percentiles for pre- and post-BMIs were recorded. The data are provided in the table below:

Gender	Age	Group	PreBMI	PostBMI	Gender	Age	Group	PreBMI	PostBMI
F	6	Cx	85.7	83.8	M	6	Cx	92.6	88.1
F	6	Cx	93.8	92.9	M	7	Cx	95.8	94.7
F	7	Cx	93.5	92.5	M	7	Cx	90.4	89.1
F	8	Cx	90.1	89.8	M	7	Cx	91.2	88.6
F	9	Tx	92.3	90.7	M	8	Tx	94.4	87.8
F	9	Tx	90.3	88.3	M	8	Tx	93.2	87.3
F	12	Cx	87.6	85.9	M	10	Cx	93.9	91.5
F	12	Cx	87.2	84.1	M	10	Tx	96.2	91.1
F	12	Tx	96.9	94.9	M	10	Tx	89.4	87.9
F	12	Tx	85.8	81.2	M	11	Tx	86.2	77.1
F	13	Cx	96.7	94.1	M	11	Tx	95.4	84.8
F	13	Cx	93.5	92.9	M	12	Cx	97.7	95.8
F	13	Tx	92.3	87.5	M	13	Tx	85.3	80.0
F	13	Tx	85.3	83.7	M	13	Tx	86.2	82.4
F	14	Tx	95.5	78.7	M	14	Cx	85.5	83.6
F	15	Cx	91.3	89.9	M	14	Cx	97.8	93.8
F	15	Tx	95.8	87.1	M	16	Cx	95.0	93.6
F	16	Tx	90.7	87.2	M	16	Tx	93.1	86.8

(a) Is the decrease in BMI percentile (preBMI-postBMI) normally distributed? Plot a histogram and test for normality of the distribution.

(b) Find the optimal lambda for Box-Cox transformation. Transform the change in BMI percentile (find the appropriate transformation in Table 2.1), and show that the transformed variable is normally distributed. Plot the histogram and do a formal testing.

(c) Fit the general regression model to the Box-Cox tranformed change in BMI percentile. Does this model have a good fit?

(d) What predictors are significant at the 5% level? Write the interpretation of the estimated regression coefficients for the significant predictors only.

(e) Predict change in BMI percentile for a 9-year old girl in the control group.

EXERCISE 2.2. Investigators at a large medical center conducted a quality improvement (QI) study which consisted of a six-month-long series of seminars and practical instructional tools on how to improve quality assurance for future projects at this center. Data were collected on participants' designation (nurse/doctor/staff), years of work at the center, whether had a prior experience with QI projects, and the score on the knowledge and attitude test taken at the end of the study. The score was constructed as the sum of 20 questions on a 5-point Likert scale, thus potentially ranging between 20 and 100. The large value indicates better knowledge about QI and more confidence and desire to use it in upcoming projects. The data on 45 study participants are summarized as follows:

Desgn	Wrkyrs	PriorQI	Score	Desgn	Wrkyrs	PriorQI	Score
nurse	16	yes	63	nurse	8	yes	62
nurse	9	yes	93	nurse	22	yes	68
nurse	8	yes	74	nurse	4	no	93
nurse	1	no	69	nurse	6	no	77
nurse	5	no	67	nurse	2	no	59
nurse	3	no	66	nurse	20	no	64
nurse	24	no	86	nurse	2	no	70
nurse	4	no	74	nurse	3	no	63
nurse	1	no	88	nurse	16	no	65
nurse	24	no	84	nurse	18	no	73
nurse	3	no	97	nurse	15	no	76
doctor	2	yes	88	doctor	2	yes	85
doctor	5	yes	78	doctor	7	yes	91
doctor	26	yes	82	doctor	2	yes	69
doctor	3	no	57	doctor	20	no	66
doctor	3	no	88	doctor	13	no	55
doctor	15	no	78	doctor	8	no	62
doctor	4	no	65	doctor	14	no	61
doctor	25	no	78	staff	9	yes	57
staff	3	yes	62	staff	11	yes	69
staff	21	no	55	staff	19	no	64
staff	8	no	62	staff	17	no	76
staff	11	no	67				

(a) Construct a histogram of the score. Does the distribution look normal? Perform the test for normality. Draw conclusion.

(b) Transform the score variable using a meaningful Box-Cox transformation and assure that it is now normally distributed by plotting the histogram and doing normality testing.

(c) Run the general linear regression model on the transformed score. What predictors are significant at the 0.05 level?

(d) Interpret the estimates of the significant beta coefficients. Does the model fit the data well? Conduct the chi-squared deviance test.

(e) Predict the score for a nurse who has worked at the center for seven years and who had previously been a co-PI on a grant that involved quality assurance component.

EXERCISE 2.3. A group of 24 beginner-level cycling enthusiasts met at the park for bicycling race that consists of biking for 30 minutes along a 1.3-mile loop.

The winner is the one who bikes the longest distance. The information recorded for each participant is gender (M/F), whether had a prior experience in races like this (yes/no), self-evaluation of abilities to finish the race and do well (on a 10-point scale, with 10 being the highest confidence), and the distance biked (in miles). The data are presented below:

Gender	Prior Expr	Self Eval	Distance	Gender	Prior Expr	Self Eval	Distance
F	no	2	1.9	F	no	7	4.4
F	no	2	2.1	F	yes	3	3.1
F	yes	8	3.8	M	yes	10	6.4
F	yes	4	3.0	F	yes	4	3.2
M	no	5	4.2	F	no	6	5.1
F	yes	10	8.2	M	no	10	5.9
F	no	3	3.1	F	no	6	5.0
F	no	4	2.4	M	yes	3	3.6
F	no	5	4.6	F	no	7	4.4
M	yes	6	8.7	M	yes	10	11.2
F	no	6	4.7	F	yes	3	3.0
M	yes	7	4.2	M	yes	7	4.3

(a) Are the distances normally distributed? Plot the histogram, do the testing. Explain.

(b) Create indicator variables `male` and `prior_yes` (existing prior experience), and use them to find a meaningful Box-Cox transformation that would transform the distance into a normally distributed variable. Prove its normality.

(c) Fit the general linear regression model to the transformed distance. Show that the model has a good fit. Discuss significance of predictors.

(d) Give interpretation for the estimates of the statistically significant regression coefficients. Use alpha=0.05.

(e) Write down the final model that can be used for prediction of distance. Predict the distance that a woman with no prior experience would bike, if she is moderately confident about her abilities with the self-assessment value of 5.

EXERCISE 2.4. A health insurance firm is doing an analysis of the aggregate insurance claims that were received in a particular fiscal year. Investigators randomly select 40 companies that are insured by this firm and, for each company, pull out the data on the number of policies, number of years insured with the firm, percent of open claims from last year, and the aggregate claim amount from this year (in millions of dollars). The data are:

Num policies	Yrs w/ firm	Open claims	Claim amount	Num policies	Yrs w/ firm	Open claims	Claim amount
12318	4	16	19.9	29629	9	35	107.4
29777	4	15	200.5	32319	6	19	78.9
36980	10	12	308.5	27103	23	25	0.3
18055	4	20	24.4	23704	2	28	6.1
16505	20	27	48.7	20432	21	16	58.4
19049	11	14	51.0	30899	16	12	19.5
37112	20	26	163.2	19052	10	23	46.9
22338	16	35	7.1	37823	12	19	325.6
32349	16	25	1.5	24269	14	31	5.7
26626	1	21	81.0	23103	22	14	71.2
28547	11	17	91.0	25556	4	32	29.3
33268	5	21	147.5	15878	11	12	34.4
29045	13	29	63.9	36772	17	13	50.6
18622	7	10	8.5	19475	1	34	107.5
22784	12	11	27.0	29241	8	29	180.2
39612	23	26	296.6	36821	7	33	158.7
28423	7	12	129.0	47309	11	12	124.0
17020	6	30	26.0	15381	2	25	41.9
36930	7	24	98.6	39857	13	11	195.0
37152	15	26	103.5	34790	7	18	60.7

(a) Plot a histogram and carry out statistical tests for normality of the distribution of claim amounts. Transform the variable via a Box-Cox transformation to achieve normality. Show that the transformed variable is normally distributed.

(b) Fit a linear regression model, relating the transformed claim amounts to all the other variables. Which variables are significant predictors at the 5% level?

(c) Assess the model fit. Interpret estimated significant regression coefficients.

(d) Compute the predicted amount of aggregate claims for a company with 15,500 policy holders, that has been buying policies at this firm for the past three years, and that still has 15% of outstanding claims from the previous year.

EXERCISE 2.5. Show that a gamma distribution with density defined by (2.3) belongs to the exponential family of distributions (1.3) with $\theta = -1/(\alpha\beta)$ and $\phi = 1/\alpha$. Conclude that the gamma regression is a generalized linear regression. Give its link function.

EXERCISE 2.6. For the data in Exercise 2.1,
(a) Fit the gamma regression model with the log link function. Write down the fitted model. Check its goodness of fit.
(b) What variables are significant predictors in this model? Use the 5% significance level.
(c) Interpret estimated significant regression coefficients.
(d) Predict change in BMI percentile for a 9-year old girl in the control group. Compare the prediction with the one obtained in Exercise 2.1.

EXERCISE 2.7. For the data in Exercise 2.2,
(a) Fit the gamma regression model with the log link function. Present the fitted model and discuss its goodness-of-fit.
(b) Discuss significance of the beta coefficients. Interpret the estimated significant coefficients.
(c) Predict the score for a nurse who has worked at the center for seven years and who had previously been a co-PI on a grant that involved quality assurance component. Compare that predicted score to the one obtained in Exercise 2.2.

EXERCISE 2.8. Consider the data set in Exercise 2.3. Fit the gamma model and do the following:
(a) Write out explicitly the estimated model. Check goodness of fit of this model.
(b) Which predictors would really influence the response, if changed? Give interpretation of the estimated significant regression coefficients.
(c) Predict the distance that a woman with no prior experience would bike, if she is moderately confident about her abilities with the self-assessment value of 5. Compare your answer to the one obtained in Exercise 2.3.

EXERCISE 2.9. Refer to Exercise 2.4. Answer the questions below for the data in that exercise.
(a) Run the gamma regression and write the predicted model. What variables are significant predictors of the claim amount? Compare to the model in Exercise 2.4.
(b) Interpret estimates of the significant beta coefficients. How good is the model fit?
(c) Obtain the predicted amount of aggregate claims for a company with 15,500 policy holders, that has been buying policies at this firm for the past three years, and that still has 15% of outstanding claims from the previous year. Compare the result with the one computed in Exercise 2.4.

Chapter 3

Regression Models for Binary Response

Suppose the response variable y is *binary* (or *dichotomous*) variable, that is, it assumes only two possible values. For simplicity, we will denote these values by 0 and 1. The relation between y and predictors cannot be modeled by a linear regression because the error terms would not be normally distributed. The way out of this predicament is to use the generalized linear regression approach and to model not y itself but the probability that y is equal to one. In this chapter, three models for the binary response are presented: logistic, probit, and complementary log-log models.

3.1 Binary Logistic Regression Model

3.1.1 Model Definition

Denote by π the probability that y is equal to one, that is, $\pi = \mathbb{P}(y = 1)$. Note that π is also the mean of y. Indeed, $\mathbb{E}(y) = (1)(\pi) + (0)(1 - \pi) = \pi$. The *binary* (or *dichotomous*) *logistic regression model*[1] with the predictors $x_1, \ldots, x_k$ has the form:

$$\pi = \mathbb{E}(y) = \frac{\exp\{\beta_0 + \beta_1 x_1 + \cdots + \beta_k x_k\}}{1 + \exp\{\beta_0 + \beta_1 x_1 + \cdots + \beta_k x_k\}}.$$

The name "logistic" comes from the fact that the distribution with the cumulative distribution function $F(x) = \dfrac{e^x}{1 + e^x}$, $-\infty < x < \infty$, is called the *logistic*

---

[1] Introduced in Cox, D.R. (1958)."The regression analysis of binary sequences". *Journal of the Royal Statistical Society, Series B*, 20(2): 215 – 242.

distribution.

An alternative form of the binary logistic regression model is derived via the use of the *logit link function* of π ("logit"="logistic"+"unit"), defined as:

$$\text{logit}\,\pi = \ln \frac{\pi}{1-\pi}\,.$$

The ratio $\dfrac{\pi}{1-\pi} = \dfrac{\mathbb{P}(y=1)}{\mathbb{P}(y=0)}$ represents the *odds in favor of* the event $y = 1$.

Using the logit transformation, the binary logistic regression model may be written in the form (verify!)

$$\text{logit}\,\pi = \ln \frac{\pi}{1-\pi} = \beta_0 + \beta_1 x_1 + \cdots + \beta_k x_k\,. \tag{3.1}$$

Thus, the binary logistic regression is a linear model for the natural logarithm of the odds and hence is sometimes called *log-odds model.*

It can be proven (see Exercise 3.1) that the logistic regression is an example of a generalized linear model with the logit link function.

3.1.2 Fitted Model

The only model parameters that are unknown and have to be estimated from the observations are the beta regression coefficients. The fitted binary logistic model has the form

$$\widehat{\pi} = \widehat{\mathbb{E}}(y) = \frac{\exp\{\widehat{\beta}_0 + \widehat{\beta}_1 x_1 + \cdots + \widehat{\beta}_k x_k\}}{1 + \exp\{\widehat{\beta}_0 + \widehat{\beta}_1 x_1 + \cdots + \widehat{\beta}_k x_k\}}\,. \tag{3.2}$$

Alternative ways to write the fitted model, which aid better in interpretation of the estimated beta coefficients, are:

$$\text{logit}\,\widehat{\pi} = \ln \frac{\widehat{\pi}}{1-\widehat{\pi}} = \widehat{\beta}_0 + \widehat{\beta}_1 x_1 + \cdots + \widehat{\beta}_k x_k, \tag{3.3}$$

and

$$\frac{\widehat{\pi}}{1-\widehat{\pi}} = \exp\{\widehat{\beta}_0 + \widehat{\beta}_1 x_1 + \cdots + \widehat{\beta}_k x_k\}. \tag{3.4}$$

3.1.3 Interpretation of Estimated Regression Coefficients

In view of (3.3) and (3.4), in the logistic regression model, the estimates of the regression coefficients yield the following interpretation.

- If a predictor variable x_1 is numeric, then the estimated regression parameter $\widehat{\beta}_1$ can be interpreted as the estimated *change in the log-odds* for every unit increase in x_1, holding all the other predictors fixed. Indeed, if x_1 is replaced by $x_1 + 1$, the difference in estimated log-odds is

$$\ln \frac{\widehat{\pi}(x_1 + 1)}{1 - \widehat{\pi}(x_1 + 1)} - \ln \frac{\widehat{\pi}(x_1)}{1 - \widehat{\pi}(x_1)}$$

$$= \widehat{\beta}_0 + \widehat{\beta}_1 (x_1 + 1) + \widehat{\beta}_2 x_2 + \cdots + \widehat{\beta}_k x_k - (\widehat{\beta}_0 + \widehat{\beta}_1 x_1 + \cdots + \widehat{\beta}_k x_k) = \widehat{\beta}_1.$$

Alternatively, the quantity $(\exp\{\widehat{\beta}_1\} - 1) \cdot 100\%$ represents the estimated *percent change in odds* when x_1 is increased by one unit, and the other predictors are held fixed. This can be seen by writing:

$$\frac{\frac{\widehat{\pi}(x_1+1)}{1-\widehat{\pi}(x_1+1)} - \frac{\widehat{\pi}(x_1)}{1-\widehat{\pi}(x_1)}}{\frac{\widehat{\pi}(x_1)}{1-\widehat{\pi}(x_1)}} \cdot 100\% = \left(\frac{\frac{\widehat{\pi}(x_1+1)}{1-\widehat{\pi}(x_1+1)}}{\frac{\widehat{\pi}(x_1)}{1-\widehat{\pi}(x_1)}} - 1 \right) \cdot 100\%$$

$$= \left(\frac{\exp\{\widehat{\beta}_0 + \widehat{\beta}_1(x_1 + 1) + \widehat{\beta}_2 x_2 + \cdots + \widehat{\beta}_k x_k\}}{\exp\{\widehat{\beta}_0 + \widehat{\beta}_1 x_1 + \cdots + \widehat{\beta}_k x_k\}} - 1 \right) \cdot 100\% = \left(\exp\{\widehat{\beta}_1\} - 1 \right) \cdot 100\% .$$

- If a predictor variable x_1 is an indicator variable, then the estimated parameter $\widehat{\beta}_1$ can be interpreted as the estimated *difference in log-odds* when $x_1 = 1$ and when $x_1 = 0$, controlling for all the other predictors. To see that, we write

$$\ln \frac{\widehat{\pi}(x_1 = 1)}{1 - \widehat{\pi}(x_1 = 1)} - \ln \frac{\widehat{\pi}(x_1 = 0)}{1 - \widehat{\pi}(x_1 = 0)}$$

$$= \widehat{\beta}_0 + \widehat{\beta}_1 \cdot 1 + \widehat{\beta}_2 x_2 + \cdots + \widehat{\beta}_k x_k - (\widehat{\beta}_0 + \widehat{\beta}_1 \cdot 0 + \widehat{\beta}_2 x_2 + \cdots + \widehat{\beta}_k x_k) = \widehat{\beta}_1.$$

Alternatively, the estimated ratio of odds for $x_1 = 1$ and that for $x_1 = 0$, expressed as percentage, is

$$\frac{\frac{\widehat{\pi}(x_1=1)}{1-\widehat{\pi}(x_1=1)}}{\frac{\widehat{\pi}(x_1=0)}{1-\widehat{\pi}(x_1=0)}} \cdot 100\% = \frac{\exp\{\widehat{\beta}_0 + \widehat{\beta}_1 \cdot 1 + \widehat{\beta}_2 x_2 + \cdots + \widehat{\beta}_k x_k\}}{\exp\{\widehat{\beta}_0 + \widehat{\beta}_1 \cdot 0 + \widehat{\beta}_2 x_2 + \cdots + \widehat{\beta}_k x_k\}} \cdot 100\%$$

$$= \exp\{\widehat{\beta}_1\} \cdot 100\%,$$

hence, the quantity $\exp\{\widehat{\beta}_1\} \cdot 100\%$ represents the estimated *percent ratio in odds* when $x_1 = 1$ and when $x_1 = 0$, while the other predictors are held constant.

3.1.4 Predicted Probability

Taking into consideration (3.2), for particular values of predictor variables $x_1^0, \ldots, x_k^0$, the predicted probability π^0 can be found as

$$\pi^0 = \frac{\exp\{\widehat{\beta}_0 + \widehat{\beta}_1 \, x_1^0 + \cdots + \widehat{\beta}_k \, x_k^0\}}{1 + \exp\{\widehat{\beta}_0 + \widehat{\beta}_1 \, x_1^0 + \cdots + \widehat{\beta}_k \, x_k^0\}}.$$

3.1.5 SAS Implementation

The procedure `genmod` with `dist=binomial` and `link=logit` may be applied to fit the binary logistic regression. SAS orders the levels of the response variable in alphabetical order (if categorical) or in ascending order (if numeric) and models the probability of the lower level. To reverse that, one has to include `descending` after `proc genmod` statement.

3.1.6 R Implementation

The function `glm()` with specifications `family=binomial(link=logit)` will fit a binary logistic regression model.

3.1.7 Example

EXAMPLE 3.1. A professor of organization and management is interested in studying the factors that influence the approach that company managers promote among their employees, competition or collaboration. The data on 50 companies are collected. The variables are the type of company ownership (sole ownership, partnership, or stock company), the number of employees, and the promoted approach (competition or collaboration). The SAS code given below models the probability of collaboration via logistic regression. In the output, we also included the values of AIC, AICC, and BIC criteria which will be utilized later, in Example 3.3, to compare fits of logistic, probit, and complementary log-log models.

```
data companies;
input ownership$ nemployees approach$ @@;
cards;
partner 88  comp  sole     60  coll  stock    24  comp
partner 108 coll  stock    88  coll  stock    119 comp
partner 25  comp  partner  53  comp  stock    82  comp
stock   69  coll  stock    24  comp  partner  94  coll
```

```
stock      86  coll  stock   46  coll  partner 106 coll
stock      92  coll  stock   22  coll  stock    94 coll
sole       20  coll  stock   30  comp  partner  92 coll
sole       90  coll  stock  114  coll  sole     66 coll
sole       26  coll  stock   59  comp  sole     54 coll
sole       69  comp  partner  75  comp  partner 62 coll
stock      26  comp  sole    112  coll  sole     87 comp
stock     104  comp  partner  25  comp  stock    63 comp
stock      21  comp  sole     93  comp  stock    91 coll
sole       33  comp  sole     97  coll  sole     93 coll
stock      41  comp  partner  73  comp  partner  32 comp
partner    56  comp  partner  98  comp  partner  97 comp
stock     117  coll  stock    75  comp
;

proc genmod;
 class ownership(ref='partner');
   model approach=ownership nemployees/
             dist=binomial link=logit;
run;
```

The relevant output is as follows:

```
Full Log Likelihood     -29.9675
AIC     67.9350
AICC    68.8239
BIC     75.5831
```

Analysis Of Maximum Likelihood Parameter Estimates

Parameter		Estimate	Pr > ChiSq
Intercept		-2.5147	0.0148
ownership	sole	1.7388	0.0455
ownership	stock	0.6726	0.3629
nemployees		0.0241	0.0267

To check how good the fit of this model is, we run the statements below.

```
proc genmod;
   model approach=/dist=binomial link=logit;
run;
```

Full Log Likelihood   -34.6173

```
data deviance_test;
 deviance=-2*(-34.6173-(-29.9675));
   pvalue=1-probchi(deviance,3);
run;

proc print;
run;
```

deviance   pvalue
 9.2996 0.025562

Since the p-value is not in excess of 0.05, we conclude that the fitted model has a better fit than the intercept-only model. The fitted model for $\pi = \mathbb{P}(collaboration)$ is

$$\text{logit } \widehat{\pi} = -2.5146 + 1.7388 * sole + 0.6726 * stock + 0.0241 * nemployees,$$

with sole ownership and number of employees being significant predictors at the 5% level of significance. In this model, the estimated odds in favor of the collaborative approach in companies with a single owner are $\exp\{1.7388\} \cdot 100\% = 569.1\%$ of those in companies with multiple partners. For every additional employee, the estimated odds of collaborative strategy increase by $\left(\exp\{0.0241\} - 1\right) \cdot 100\% = 2.4\%$.

Finally, suppose the professor would like to estimate the probability of the collaborative approach in a solely owned company with 40 employees. To predict this probability we compute

$$\mathbb{P}^0(collaboration) = \frac{\exp\{-2.5146 + 1.7388 + 0.0241 * 40\}}{1 + \exp\{-2.5146 + 1.7388 + 0.0241 * 40\}} = 0.5469.$$

To get the same predicted probability in SAS, we run the following code:

```
data prediction;
input ownership$ nemployees;
cards;
sole 40
;

data companies;
 set companies prediction;
run;
```

```
proc genmod data=companies;
 class ownership (ref='partner');
   model approach =ownership nemployees/
       dist=binomial link=logit;
 output out=outdata p=pred_probcoll;
run;

proc print data=outdata (firstobs=51 obs=51);
var pred_probcoll;
run;
```

pred_probcoll
     0.54688

The R output presented below matches the output of the SAS code.

```
companies.data<- read.csv(file="./Example3.1Data.csv",
header=TRUE, sep=",")

#specifying reference categories
ownership.rel<- relevel(companies.data$ownership, ref="partner")
approach.rel<- relevel(companies.data$approach, ref="comp")

#fitting logistic model
summary(fitted.model<- glm(approach.rel ~ ownership.rel
+ nemployees,
data=companies.data, family=binomial(link=logit)))
```

Coefficients:

	Estimate	Pr(>|z|)
(Intercept)	-2.51469	0.0148
ownership.relsole	1.73882	0.0455
ownership.relstock	0.67256	0.3629
nemployees	0.02410	0.0267

AIC: 67.935

```
#extracting AICC and BIC for fitted model
p<- 4
n<- 50
print(AICC<- -2*logLik(fitted.model)+2*p*n/(n-p-1))
```

68.82389

```
BIC(fitted.model)
```

75.58309

```
#checking model fit
intercept.only.model<- glm(approach.rel ~ 1,
family=binomial(link=logit))
print(deviance<- -2*(logLik(intercept.only.model)
-logLik(fitted.model)))
```

9.2997

```
print(p.value<- pchisq(deviance, df=3, lower.tail=FALSE))
```

0.02556052

```
#using fitted model for prediction
print(predict(fitted.model, type="response",
data.frame(ownership.rel="sole", nemployees=40)))
```

0.5468756

□

3.2 Probit Model

3.2.1 Model Definition

Consider a binary response variable y and predictors $x_1, \ldots, x_k$. Let $\pi = \mathbb{P}(y = 1) = \mathbb{E}(y)$. The *probit regression model*[2] ("probit"="probability"+"unit") is given by:

$$\pi = \mathbb{E}(y) = \Phi\big(\beta_0 + \beta_1\, x_1 + \cdots + \beta_k\, x_k\big)$$

where $\Phi(\cdot)$ is the cumulative distribution function of a $\mathcal{N}(0,1)$ distribution.

Alternatively, the above relation may be written as

$$\Phi^{-1}(\pi) = \Phi^{-1}(\mathbb{E}(y)) = \beta_0 + \beta_1\, x_1 + \cdots + \beta_k\, x_k.$$

The function $\Phi^{-1}(\cdot)$ is called the *probit link function*.

It may be shown (see Exercise 3.1) that the probit regression model belongs to the class of generalized linear models.

---

[2]First appears in Bliss, C. I. (1935). "The calculation of the dosage-mortality curve". *Annals of Applied Biology*, 22(1): 134 – 167.

3.2.2 Fitted Model

The fitted probit model looks like this:

$$\Phi^{-1}(\widehat{\pi}) = \widehat{\beta}_0 + \widehat{\beta}_1 \, x_1 + \cdots + \widehat{\beta}_k \, x_k, \tag{3.5}$$

or, equivalently,

$$\widehat{\pi} = \Phi\big(\widehat{\beta}_0 + \widehat{\beta}_1 \, x_1 + \cdots + \widehat{\beta}_k \, x_k\big). \tag{3.6}$$

3.2.3 Interpretation of Estimated Regression Coefficients

Taking into account (3.5), estimated regression coefficients in the probit model are interpreted in terms of z-score (or *probit index*), which is a percentile of the standard normal distribution.

• For a numeric predictor variable x_1, the estimated coefficient $\widehat{\beta}_1$ gives the change in the estimated z-score of the response variable for a unit increase in x_1, keeping all the other predictors unchanged. Indeed, we have

$$\Phi^{-1}\big(\widehat{\pi}(x_1 + 1)\big) - \Phi^{-1}\big(\widehat{\pi}(x_1)\big) = \widehat{\beta}_0 + \widehat{\beta}_1(x_1 + 1) + \widehat{\beta}_2 \, x_2 + \cdots + \widehat{\beta}_k \, x_k$$
$$- \big(\widehat{\beta}_0 + \widehat{\beta}_1 \, x_1 + \cdots + \widehat{\beta}_k \, x_k\big) = \widehat{\beta}_1.$$

• For a 0-1 predictor x_1, the estimated coefficient $\widehat{\beta}_1$ is interpreted as a difference in the estimated z-scores of the response variable for the levels $x_1 = 1$ and $x_1 = 0$, controlling for the other predictors, since

$$\Phi^{-1}\big(\widehat{\pi}(x_1 = 1)\big) - \Phi^{-1}\big(\widehat{\pi}(x_1 = 0)\big) = \widehat{\beta}_0 + \widehat{\beta}_1 \cdot 1 + \widehat{\beta}_2 \, x_2 + \cdots + \widehat{\beta}_k \, x_k$$
$$- \big(\widehat{\beta}_0 + \widehat{\beta}_1 \cdot 0 + \widehat{\beta}_2 \, x_2 + \cdots + \widehat{\beta}_k \, x_k\big) = \widehat{\beta}_1.$$

3.2.4 Predicted Probability

Suppose we are given certain values of the predictors $x_1^0, \ldots, x_k^0$. Then, in accordance with (3.6), we predict the probability π^0 by

$$\pi^0 = \Phi\big(\widehat{\beta}_0 + \widehat{\beta}_1 \, x_1^0 + \cdots + \widehat{\beta}_k \, x_k^0\big).$$

3.2.5 SAS Implementation

To fit a probit model, use the procedure `genmod` with the option `dist=binomial` `link=probit`.

3.2.6 R Implementation

The function glm() with family=binomial(link=probit) may be used to fit a probit model in R.

3.2.7 Example

EXAMPLE 3.2. We fit the probit model to the data in Example 3.1. The code and relevant output are given below.

```
proc genmod;
  class ownership(ref='partner');
    model approach=ownership nemployees/
      dist=binomial link=probit;
run;
```

Full Log Likelihood   -29.9519
AIC   67.9038
AICC   68.7926
BIC   75.5518

Analysis Of Maximum Likelihood Parameter Estimates

Parameter		Estimate	Pr > ChiSq
Intercept		-1.5544	0.0099
ownership	sole	1.0584	0.0404
ownership	stock	0.4227	0.3449
nemployees		0.0148	0.0214

```
proc genmod;
  class ownership(ref='partner');
    model approach=/dist=binomial link=probit;
run;
```

Full Log Likelihood   -34.6173

```
data deviance_test;
  deviance=-2*(-34.6173 -(-29.9519));
    pvalue=1-probchi(deviance,3);
run;

proc print;
run;
```

```
deviance    pvalue
 9.3308  0.025201
```

The p-value for the deviance test is less than 0.05, indicating a good fit of the model with all predictors. At the 5% significance level, sole ownership and the number of employees are significant predictors, as in the logistic model. The fitted probit model is

$$\Phi^{-1}\left(\widehat{\pi}\right) = -1.5544 + 1.0584 * sole + 0.4227 * stock + 0.0148 * nemployees.$$

The estimated z-score of the collaborative approach in companies with a single owner is larger by 1.0584 than that in companies with multiple partners. For every additional employee, the estimated z-score of the collaborative strategy increases by 0.0148.

To estimate the probability of the collaborative approach in a solely owned company with 40 employees, we calculate

$$\pi^0 = \Phi(-1.5544 + 1.0584 + 0.0148 * 40) = 0.5382.$$

SAS gives a similar prediction. The code follows.

```
data prediction;
  input ownership$ nemployees;
cards;
sole 40
;

data companies;
 set companies prediction;
run;

proc genmod;
 class ownership(ref='partner');
   model approach =ownership nemployees/
          dist=binomial link=probit;
 output out=outdata p=pred_probcoll;
run;

proc print data=outdata (firstobs=51 obs=51);
var pred_probcoll;
run;
```

```
pred_probcoll
    0.53774
```

The R script which output matches the output in SAS is as follows:

```
companies.data<-read.csv(file="./Example3.1Data.csv",
header=TRUE, sep=",")

# specifying reference categories
ownership.rel<- relevel(companies.data$ownership,ref="partner")
approach.rel<- relevel(companies.data$approach, ref = "comp")

# fitting probit model
summary(fitted.model <- glm(approach.rel ~ ownership.rel
+ nemployees,
data=companies.data, family=binomial(link=probit)))
```

```
Coefficients:
                    Estimate   Pr(>|z|)
(Intercept)        -1.554466    0.0099
ownership.relsole   1.058424    0.0404
ownership.relstock  0.422743    0.3449
nemployees          0.014770    0.0214
```

AIC: 67.904

```
#extracting AICC and BIC for fitted model
p<- 4
n<- 50
print(AICC<- -2*logLik(fitted.model)+2*p*n/(n-p-1))
```

68.79264

```
BIC(fitted.model)
```

75.55185

```
#checking model fit
intercept.only.model<- glm(approach.rel ~ 1,
family=binomial(link=probit))
print(deviance<- -2*(logLik(intercept.only.model)
-logLik(fitted.model)))
```

9.330941

```
print(p.value<- pchisq(deviance, df=3, lower.tail=FALSE))
```

0.02519957

```
#using fitted model for prediction
print(predict(fitted.model, type="response",
data.frame(ownership.rel="sole", nemployees=40)))
```

0.5377392

□

3.3 Complementary Log-Log Model

3.3.1 Model Definition

Consider a binary response variable y and predictors $x_1, \ldots, x_k$. Let $\pi = \mathbb{P}(y = 1)$. The *complementary log-log model*[3] has the form

$$\pi = 1 - \exp\left\{ - \exp\left\{ \beta_0 + \beta_1 x_1 + \cdots + \beta_k x_k \right\} \right\},$$

or, equivalently,

$$\ln\left(- \ln(1 - \pi) \right) = \beta_0 + \beta_1 x_1 + \cdots + \beta_k x_k.$$

Here the function $\ln\left(- \ln(1 - \pi) \right)$ is referred to as *complementary log-log link function*, since the log-log function is applied to the complement of π. On a side note, the function $F(x) = \exp\{ - \exp\{-x\} \}$, $-\infty < x < \infty$, that is used in this model, is the cumulative distribution function of the *Gumbel* or *extreme value* distribution.

Similar to the binary logistic and probit models, the complementary log-log regression represents the class of generalized linear models (see Exercise 3.1).

3.3.2 Fitted Model

To ease interpretation of the estimated regression coefficients (see below), it is convenient to write the fitted model in the form

$$1 - \widehat{\pi} = \exp\left\{ - \exp\left\{ \widehat{\beta}_0 + \widehat{\beta}_1 x_1 + \cdots + \widehat{\beta}_k x_k \right\} \right\}. \tag{3.7}$$

---

[3] First application of this model appeared in Draper, C. C., Voller, A., and R. G. Carpenter, (1972). "The Epidemiologic interpretation of serologic data in malaria". *American Journal of Tropical Medicine and Hygiene*, 21 (5 Suppl): 696 – 703.

3.3.3 Interpretation of Estimated Regression Coefficients

By (3.7), in the complementary log-log model, the estimated regression coefficients are interpreted as follows.

- For a numeric random predictor x_1, the estimate of the coefficient $\widehat{\beta}_1$ gives the estimated rate ratio for a unit increase in x_1. To see that, we write

$$1 - \widehat{\pi}(x_1 + 1) = \exp\left\{-\exp\left\{\widehat{\beta}_0 + \widehat{\beta}_1(x_1 + 1) + \widehat{\beta}_2 x_2 + \cdots + \widehat{\beta}_k x_k\right\}\right\}$$

$$= \exp\left\{-\exp\left\{\widehat{\beta}_0 + \widehat{\beta}_1 x_1 + \cdots + \widehat{\beta}_k x_k\right\}\exp\{\widehat{\beta}_1\}\right\} = \left(1 - \widehat{\pi}(x_1)\right)^{\exp\{\widehat{\beta}_1\}},$$

that is, the probability $1 - \widehat{\pi}(x_1 + 1)$ equals the probability $1 - \widehat{\pi}(x_1)$ raised to the power $\exp\{\widehat{\beta}_1\}$.

- If x_1 is an indicator variable, then the estimated regression coefficient $\widehat{\beta}_1$ is interpreted as the estimated rate ratio for $x_1 = 1$ and $x_1 = 0$. Indeed, we have

$$1 - \widehat{\pi}(x_1 = 1) = \exp\left\{-\exp\left\{\widehat{\beta}_0 + \widehat{\beta}_1 \cdot 1 + \widehat{\beta}_2 x_2 + \cdots + \widehat{\beta}_k x_k\right\}\right\}$$

$$= \exp\left\{-\exp\left\{\widehat{\beta}_0 + \widehat{\beta}_1 \cdot 0 + \widehat{\beta}_2 x_2 + \cdots + \widehat{\beta}_k x_k\right\}\exp\{\widehat{\beta}_1\}\right\} = \left(1 - \widehat{\pi}(x_1 = 0)\right)^{\exp\{\widehat{\beta}_1\}},$$

that is, the probability $1 - \widehat{\pi}(x_1 = 1)$ equals to the probability $1 - \widehat{\pi}(x_1 = 0)$ raised to the power $\exp\{\widehat{\beta}_1\}$.

3.3.4 Predicted Probability

Using the expression (3.7), for a fixed set of predictors $x_1^0, \ldots, x_k^0$, the predicted probability π_0 is calculated as

$$\pi^0 = 1 - \exp\left\{-\exp\left\{\widehat{\beta}_0 + \widehat{\beta}_1 x_1^0 + \cdots + \widehat{\beta}_k x_k^0\right\}\right\}.$$

3.3.5 SAS Implementation

To fit a complementary log-log model, use the procedure **genmod** with the option **dist=binomial link=cloglog** in the **model** statement.

3.3.6 R Implementation

The function glm() with family=binomial(link=cloglog) will fit a complementary log-log model.

3.3.7 Example

EXAMPLE 3.3. We fit the complementary log-log model to the data in Example 3.1. The SAS code and important output are as follows:

```
proc genmod;
  class ownership(ref='partner');
    model approach=ownership nemployees/
        dist=binomial link=cloglog;
run;
```

```
Full Log Likelihood   -30.3467
AIC     68.6934
AICC    69.5823
BIC     76.3415
```

Parameter		Estimate	Pr > ChiSq
Intercept		-2.0826	0.0058
ownership	sole	1.0631	0.0656
ownership	stock	0.4455	0.4249
nemployees		0.0161	0.0362

```
proc genmod;
  class ownership(ref='partner');
    model approach=/dist=binomial link=cloglog;
run;
```

```
Full Log Likelihood   -34.6173
```

```
data deviance_test;
  deviance=-2*(-34.6173-(-30.3467));
    pvalue=1-probchi(deviance,3);
run;
```

```
proc print;
run;
```

```
deviance     pvalue
 8.5412  0.036056
```

From the output, the estimated complement log-log model is $1 - \widehat{\pi} = 1 - \widehat{\mathbb{P}}(collaboration) = \widehat{\mathbb{P}}(competition) = \exp\left\{-\exp\left\{-2.08025 + 1.0631 * sole + 0.4455 * stock + 0.0161 * nemployees\right\}\right\}$. The p-value for the deviance test is less than 0.05, leading us to the conclusion that this model has a good fit. It is more convenient to interpret the estimated regression coefficients in terms of the probability of competition rather than collaboration. Unlike in the two previous models, sole ownership is significant here only at the 10% level, but not 5%. The number of employees is still a significant predictor at the 5% level, since their p-values are smaller than 0.05. The estimated probability of competition for single owner companies equals to the estimated probability of competition for companies owned by partners raised to the power $\exp\{1.0631\} = 2.8953$. Also, if the number of employees of a company is increased by one, the new estimated probability of competition is the old one raised to the power $\exp\{0.0161\} = 1.0162$.

The model estimates the probability of the collaborative approach in a solely owned company with 40 employees as

$$\mathbb{P}^0(collaboration) = 1 - \exp\left\{-\exp\left\{-2.08025 + 1.0631 + 0.0161 * 40\right\}\right\} = 0.4977.$$

The following statements produce the output in SAS.

```
data prediction;
  input ownership$ nemployees;
cards;
sole 40
;

data companies;
 set companies prediction;
run;

proc genmod;
  class ownership(ref='partner');
    model approach =ownership nemployees/
          dist=binomial link=cloglog;
  output out=outdata p=pred_probcoll;
run;

proc print data=outdata (firstobs=51 obs=51);
var pred_probcoll;
run;
```

pred_probcoll
        0.49708

The corresponding R script is given below. Its output matches the one by SAS.

```
companies.data<-read.csv(file="./Example3.1Data.csv",
header=TRUE, sep=",")

#specifying reference categories
ownership.rel<- relevel(companies.data$ownership,
ref="partner")
approach.rel<- relevel(companies.data$approach, ref = "comp")

#fitting complementary log-log model
summary(fitted.model <- glm(approach.rel ~ ownership.rel
+ nemployees,
data=companies.data, family=binomial(link=cloglog)))
```

Coefficients:

	Estimate	Pr(>|z|)
(Intercept)	-2.082608	0.00628
ownership.relsole	1.063128	0.06688
ownership.relstock	0.445539	0.41893
nemployees	0.016113	0.03474

AIC: 68.693

```
#extracting AICC and BIC for fitted model
p<- 4
n<- 50
print(AICC<- -2*logLik(fitted.model)+2*p*n/(n-p-1))
```

69.58228

```
BIC(fitted.model)
```

76.34148

```
#checking model fit
intercept.only.model<- glm(approach.rel ~ 1,
family=binomial(link=cloglog))
print(deviance<- -2*(logLik(intercept.only.model)
-logLik(fitted.model)))
```

8.54131

```
print(p.value<- pchisq(deviance, df=3, lower.tail=FALSE))
```

0.03605396

```
#using fitted model for prediction
print(predict(fitted.model, type="response",
data.frame(ownership.rel="sole", nemployees=40)))
```

0.4970774

Finally, to compare the fitted logistic, probit, and complementary log-log models in terms of goodness-of-fit, we compare the values for the AIC, AICC, and BIC criteria. These were displayed in examples above. For convenience, we repeat the quantities here:

	logistic	probit	cloglog
AIC	67.9350	67.9038	68.6934
AICC	68.8239	68.7926	69.5823
BIC	75.5831	75.5518	76.3415

The probit model has the smallest values in all the three criteria. We conclude that the probit model should be accorded a preference over the other two models. □

Exercises for Chapter 3

EXERCISE 3.1. Let y be a Bernoulli random variable with the probability of a success π. Show that the probability mass function belongs to the exponential family of distributions, that is, show that it can be written in the form (1.3) with the location parameter $\theta = \ln \frac{\pi}{1 - \pi}$ and the dispersion parameter $\phi = 1$. Conclude that the logistic, probit, and complement log-log models are special cases of the generalized linear regression. Specify the respective link functions.

EXERCISE 3.2. Dermatologists in a hospital study patients with acute psoriasis, a skin disease. They would like to know whether medication A is more effective in relieving the symptoms of psoriasis than medication B. The data are retrospectively collected on 30 patients. The variables are gender (M/F), age (in years), medication(A/B), and response (1=relief, 0=no relief). The data are as follows.

Gender	Age	Medication	Response	Gender	Age	Medication	Response
M	37	A	1	M	16	A	1
F	24	A	1	F	33	A	1
F	15	A	1	F	28	A	0
M	31	B	1	M	51	B	1
F	39	B	1	F	35	B	0
M	31	B	1	M	16	B	0
M	20	A	1	F	25	A	0
M	32	A	1	M	18	A	1
M	30	A	1	F	19	A	1
F	24	B	0	M	39	B	1
M	17	B	0	M	38	B	1
F	33	B	1	M	37	B	1
M	24	A	1	F	24	B	0
M	32	A	1	F	39	B	0
F	27	A	1	F	33	B	0

(a) Fit a binary logistic model. Write down the fitted model. Discuss significance of predictor variables, and goodness of fit of the model. Use $\alpha = 0.05$.

(b) Give interpretation of the estimated significant regression coefficients.

(c) Find the predicted probability of relief from psoriasis for a 50-year old woman who is administered the medication A treatment.

(d) Repeat parts (a)-(c) but fit a probit model. Compare the results.

(e) Repeat parts (a)-(c), fitting a complementary log-log model. Compare the results with the previous two models. Which of the three models has a better fit?

EXERCISE 3.3. A study is conducted to reveal what factors underlie a financial success of a novel. A random sample of 44 published novels is drawn. It is recorded whether each novel is successful financially (yes/no), whether the book cover is catchy (yes/no), the number of publisher's promotional methods (none/one/many), author's popularity (first novel/several novels/many novels), and the number of years the publisher was in business before the novel was published. The observations are

Success	Cover	Methods	Novels	Years	Success	Cover	Methods	Novels	Years
yes	yes	one	many	18	no	no	one	several	9
no	no	one	first	7	yes	no	many	several	13
no	yes	none	several	10	yes	yes	none	first	6
yes	yes	many	many	6	no	no	none	many	2
no	yes	none	several	1	yes	yes	one	several	7
no	no	one	several	1	yes	yes	many	many	17
no	no	one	first	11	yes	yes	many	first	18
yes	no	one	several	19	yes	yes	one	several	17
yes	yes	none	first	5	no	yes	none	several	9
no	no	none	many	2	no	no	one	several	11
no	no	one	several	10	yes	yes	many	first	17
no	no	many	many	9	no	no	many	many	1
yes	no	many	several	6	no	no	many	many	6
yes	yes	many	many	8	no	yes	none	several	1
no	no	one	several	12	yes	yes	many	first	6
no	no	none	many	2	yes	yes	one	many	4
yes	no	none	several	17	no	yes	none	many	7
yes	yes	many	first	10	no	no	one	first	12
yes	no	none	several	7	no	no	one	several	7
no	no	one	first	12	yes	yes	one	several	9
no	yes	none	several	7	no	no	one	several	8
no	yes	none	many	4	no	no	one	several	2

(a) Fit a binary logistic model to the data. What predictors turn out to be significant at the 5% level? How good is the fit of the model?

(b) Give interpretation of the estimated significant beta coefficients.

(c) Suppose a newly established publishing house prints a novel by some previously unknown author, and doesn't advertise the publication. Find the estimated probability that this novel is successful financially, if it has an extremely catchy cover.

(d) Redo parts (a) through (c), fitting a probit model.

(e) Redo parts (a) through (c) with the complementary log-log model. How good is the model fit compared to the logistic and probit models?

EXERCISE 3.4. A bank needs to estimate the default rate of customers' home equity loans. A random sample of 35 customers is drawn. The selected variables are loan-to-value (LTV) ratio defined as the ratio of a loan to the value of an asset purchased (in percent), age (in years), income (high/low), and response (yes=default, no=payoff). The data are

LTV	Age	Income	Default	LTV	Age	Income	Default
70	41	low	no	40	44	low	no
70	25	high	yes	80	36	high	no
65	48	low	no	90	47	high	no
65	48	high	no	80	29	high	no
60	32	high	yes	70	24	low	yes
50	48	high	no	30	42	high	no
55	53	low	no	50	33	low	no
85	38	high	yes	80	36	low	no
80	43	low	yes	75	54	low	no
50	33	low	no	75	29	high	yes
60	42	low	no	70	38	low	no
90	23	low	yes	60	35	low	no
80	31	high	no	95	30	low	yes
70	37	high	no	80	34	low	yes
40	39	high	no	75	43	low	yes
80	40	low	no	75	47	high	no
70	52	high	no	85	47	low	yes
80	29	low	yes				

(a) Run the binary logistic model, regressing on all the predictors. Identify variables that are significant predictors of loan default at the 5% level of significance. Analyze the model fit.

(b) Interpret the estimated significant beta coefficients. What is your suggestion in order for the bank to decrease the default rate of home equity loans?

(c) Give a point estimate for the probability of loan default if LTV ratio is 50%, and the borrower is a 50-year old men with high income.

(d) Repeat the previous parts, fitting a probit model. How different are the results?

(e) Redo parts (a)-(c) with a complementary log-log model. Discuss differences between the three models, if any. Which model fits the data the best?

EXERCISE 3.5. There are three basic dermatoglyphic patterns occurring in the population: loops, whorls, and arches. A cardiologist is interested in finding whether a significant difference exists in the count of the fingerprint patterns in cardiac patients and in the control group. He obtained a clear fingerprint pattern readings for 24 male cardiac patients, and those for 24 healthy males. The variables are group (1=cardiac, 0=control), the number of arches (A), and the number of whorls (W). Note that the number of loops (L) can be recovered by subtraction from 10, therefore, it is not included as a predictor. The data are presented below.

Group	A	W	Group	A	W	Group	A	W	Group	A	W
1	8	2	1	2	7	0	0	9	0	3	2
1	1	2	1	2	8	0	2	1	0	1	2
1	2	1	1	6	0	0	0	8	0	2	5
1	4	0	1	3	5	0	1	3	0	4	0
1	2	7	1	1	0	0	3	1	0	8	1
1	6	3	1	7	1	0	1	4	0	2	7
1	2	8	1	4	3	0	0	8	0	0	10
1	1	9	1	2	4	0	1	6	0	0	5
1	3	0	1	5	3	0	0	9	0	0	6
1	0	2	1	7	1	0	2	2	0	2	1
1	3	2	1	8	1	0	4	4	0	0	7
1	2	7	1	0	6	0	0	6	0	0	6

(a) Model the probability of being a cardiac patient via the binary logistic regression. Write the fitted model explicitly. Discuss goodness-of-fit of the model and significance of regression coefficients. Assume $\alpha = 0.01$ for all tests.

(b) Interpret the estimated significant regression coefficients. For which fingerprint pattern the fitted probability is the largest? For which, the lowest?

(c) Suppose the model is used to predict the probability of being a cardiac patient in a male with the dermatoglyphics reading L-L-W-W-A-W-A-L-L-W. What is this predicted probability?

(d) In parts (a)-(c), fit a probit model. Compare results.

(e) Fit the complementary log-log model instead of the logistic model in (a) through (c). Do the models differ? Which of the three models should be preferred?

Chapter 4

Regression Models for Categorical Response

A natural extension of the binary (dichotomous) regression model introduced in Chapter 3 is when the categorical response variable assumes more than two values, e.g., 0, 1, or 2. This model is called a *multinomial* (or *polytomous*) logistic regression model.

Two types of multinomial logistic regression models are distinguished, depending on whether the response variable is measured on an ordinal or nominal scale. A categorical variable is measured on an *ordinal scale* (called an *ordinal variable*) if the categories have a natural ordering. For example, health status has ordered categories "poor," "fair," "good," or "excellent".

On the other hand, if the categories don't have a natural ordering, the variable is called the *nominal variable*, and is measured on *nominal scale*. In this case the categories may be simply treated as names. For instance, race is a nominal variable with unordered categories "African American," "Hispanic," "Caucasian," "Asian," and "Other".

In this chapter, we study three models for an ordinal response variable: cumulative logit, cumulative probit, and cumulative complementary log-log models; and one model for a nominal response, a generalized logit model.

4.1 Cumulative Logit Model

4.1.1 Model Definition

Let y denote an ordinal response variable with categories $1, \ldots, c$. The probability $\mathbb{P}(y \leq j)$ is the probability of y falling in one of the categories $1, \ldots, j$. This is called a *cumulative probability*. For example, if $c = 3$, the cumulative probabilities are

$$\mathbb{P}(y \leq 1) = \mathbb{P}(y = 1), \quad \mathbb{P}(y \leq 2) = \mathbb{P}(y = 1) + \mathbb{P}(y = 2),$$

and

$$\mathbb{P}(y \leq 3) = \mathbb{P}(y = 1) + \mathbb{P}(y = 2) + \mathbb{P}(y = 3) = 1.$$

The ratio $\dfrac{\mathbb{P}(y \leq j)}{\mathbb{P}(y > j)} = \dfrac{\mathbb{P}(y \leq j)}{1 - \mathbb{P}(y \leq j)}$ represents the odds of response y being in the category j or below. The *logits of the cumulative probabilities* (called *cumulative logits*) are the functions

$$\operatorname{logit} \mathbb{P}(y \leq j) = \ln \frac{\mathbb{P}(y \leq j)}{\mathbb{P}(y > j)} = \ln \frac{\mathbb{P}(y \leq j)}{1 - \mathbb{P}(y \leq j)}, \quad j = 1, \ldots, c - 1.$$

For example, with $c = 3$,

$$\operatorname{logit} \mathbb{P}(y \leq 1) = \ln \frac{\mathbb{P}(y \leq 1)}{\mathbb{P}(y > 1)} = \ln \frac{\mathbb{P}(y = 1)}{\mathbb{P}(y = 2) + \mathbb{P}(y = 3)},$$

$$\operatorname{logit} \mathbb{P}(y \leq 2) = \ln \frac{\mathbb{P}(y \leq 2)}{\mathbb{P}(y > 2)} = \ln \frac{\mathbb{P}(y = 1) + \mathbb{P}(y = 2)}{\mathbb{P}(y = 3)},$$

and since $\mathbb{P}(y \leq 3) = 1$, $\operatorname{logit} \mathbb{P}(y \leq 3)$ is not defined.

The *cumulative logit model*[1] for an ordinal response y and predictors $x_1, \ldots, x_k$ has the form

$$\operatorname{logit} \mathbb{P}(y \leq j) = \alpha_j + \beta_1 x_1 + \cdots + \beta_k x_k, \quad j = 1, \ldots, c - 1. \tag{4.1}$$

Note that this model requires a separate intercept parameter α_j for each cumulative probability. Thus, there are a total of $c + k - 1$ unknown parameters in this model ($c - 1$ intercepts and k beta coefficients). Because the beta coefficients don't depend on j, this model is sometimes termed the *proportional odds model*.

To show that the cumulative logit model belongs to the class of generalized linear models, we notice that $\mathbb{P}(y \leq j)$ is the mean of the indicator random

---

[1]Introduced in McCullagh, P. (1980). "Regression models for ordinal data". *Journal of the Royal Statistical Society, Series B*, 42 (2): 109 – 142.

variable of the event $\{y \le j\}$, which has a Bernoulli distribution. As shown in Exercise 3.1, Bernoulli distribution belongs to the exponential family. In addition, by (4.1), the link function in this model is logit.

4.1.2 Fitted Model

In view of (4.1), the fitted cumulative logit model is of the form:

$$\text{logit }\widehat{\mathbb{P}}(y \le j) = \ln \frac{\widehat{\mathbb{P}}(y \le j)}{1 - \widehat{\mathbb{P}}(y \le j)} = \widehat{\alpha}_j + \widehat{\beta}_1\, x_1 + \cdots + \widehat{\beta}_k\, x_k, \quad j = 1, \ldots, c-1,$$

or, equivalently,

$$\widehat{\mathbb{P}}(y \le j) = \frac{\exp\left\{\widehat{\alpha}_j + \widehat{\beta}_1\, x_1 + \cdots + \widehat{\beta}_k\, x_k\right\}}{1 + \exp\left\{\widehat{\alpha}_j + \widehat{\beta}_1\, x_1 + \cdots + \widehat{\beta}_k\, x_k\right\}}, \quad j = 1, \ldots, c-1. \tag{4.2}$$

4.1.3 Interpretation of Estimated Regression Coefficients

The estimates of the regression coefficients are interpreted as follows:

• If a predictor variable x_1 is numeric, then $(\exp\{\widehat{\beta}_1\} - 1) \cdot 100\%$ represents the estimated percent change in odds of the event $\{y \le j\}$ as opposed to the event $\{y > j\}$, for any $j = 1, \ldots, c-1$, when x_1 is increased by one unit, and the other predictors are held fixed. This can be seen by writing

$$\frac{\widehat{\mathbb{P}}(y \le j|x_1+1)/\widehat{\mathbb{P}}(y > j|x_1+1) - \widehat{\mathbb{P}}(y \le j|x_1)/\widehat{\mathbb{P}}(j > y|x_1)}{\widehat{\mathbb{P}}(y \le j|x_1)/\widehat{\mathbb{P}}(y > j|x_1)} \cdot 100\%$$

$$= \left(\frac{\exp\{\widehat{\alpha}_j + \widehat{\beta}_1\,(x_1+1) + \widehat{\beta}_2\, x_2 + \cdots + \widehat{\beta}_k\, x_k\}}{\exp\{\widehat{\alpha}_j + \widehat{\beta}_1\, x_1 + \widehat{\beta}_2\, x_2 + \cdots + \widehat{\beta}_k\, x_k\}} - 1\right) \cdot 100\% = (\exp\{\widehat{\beta}_1\} - 1) \cdot 100\%.$$

• If a predictor variable x_1 is an indicator variable, then the quantity $\exp\{\widehat{\beta}_1\} \cdot 100\%$ is interpreted as the ratio of odds in favor of $\{y \le j\}$ against $\{y > j\}$ when $x_1 = 1$ and when $x_1 = 0$, provided the other predictors are unchanged. This is obtained as follows:

$$\frac{\widehat{\mathbb{P}}(y \le j|x_1 = 1)/\widehat{\mathbb{P}}(y > j|x_1 = 1)}{\widehat{\mathbb{P}}(y \le j|x_1 = 0)/\widehat{\mathbb{P}}(y > j|x_1 = 0)} \cdot 100\%$$

$$= \frac{\exp\{\widehat{\alpha}_j + \widehat{\beta}_1 \cdot 1 + \widehat{\beta}_2\, x_2 + \cdots + \widehat{\beta}_k\, x_k\}}{\exp\{\widehat{\alpha}_j + \widehat{\beta}_1 \cdot 0 + \widehat{\beta}_2\, x_2 + \cdots + \widehat{\beta}_k\, x_k\}} \cdot 100\% = \exp\{\widehat{\beta}_1\} \cdot 100\%.$$

4.1.4 Predicted Probabilities

From (4.2), for given values of predictors $x_1^0, \ldots, x_k^0$, the predicted cumulative probabilities may be found as

$$\mathbb{P}^0(y \leq j) = \frac{\exp\left\{\widehat{\alpha}_j + \widehat{\beta}_1 x_1^0 + \cdots + \widehat{\beta}_k x_k^0\right\}}{1 + \exp\left\{\widehat{\alpha}_j + \widehat{\beta}_1 x_1^0 + \cdots + \widehat{\beta}_k x_k^0\right\}}, \quad j = 1, \ldots, c - 1.$$

Moreover, we can predict the probability of exact equality to a value as the difference of two consecutive predicted cumulative probabilities. Like this:

$$\mathbb{P}^0(y = j) = \mathbb{P}^0(y \leq j) - \mathbb{P}^0(y \leq j - 1), \quad j = 2, \ldots, c - 1,$$
$$\mathbb{P}^0(y = 1) = \mathbb{P}^0(y \leq 1), \quad \text{and} \quad \mathbb{P}^0(y = c) = 1 - \mathbb{P}^0(y \leq c - 1). \tag{4.3}$$

4.1.5 SAS Implementation

The procedure `genmod` with `dist=multnomial link=cumlogit` may be used to run the cumulative logit model. Since the values of the response variable must be on ordinal scale, it is convenient to present them as numbers, for instance, $1, 2, \ldots, c$, and possibly describe the levels in a `proc format` statement like this:

```
proc format;
    value response_namefmt 1='level1_name'
            2='level2_name' ... c ='levelc_name';
run;
```

Another possibility would be to enter level names as words but in `proc format` put digits 1 through c in front on the names. For example, `'1poor'`, `'2fair'`, `'3good'`, `'4excellent'`.

• By default, SAS orders the levels in alphabetical order and models the cumulative probabilities of the response variable having lower ordered values. If it is desired to reverse the order, the option `descending` may be used in the `proc genmod` statement.
• SAS predicts cumulative probability for each category, that is, the probability of being in a given category or in any category that precedes it.

4.1.6 R Implementation

Function `clm()` in the library `ordinal` may be utilized to fit the cumulative logit model. The name stands for "cumulative link models". The syntax is

74

```
summary(fitted.model.name<- clm(response.name ~ x1.name + ···
+ xk.name, data = data.name, link="logit"))
```

- R outputs beta coefficients with reversed signs. The signs of the intercepts are not changed. This means that R outputs estimates for $\alpha_1, \ldots, \alpha_{c-1}$, and $\beta_0, \ldots, \beta_k$ where

$$\text{logit}\,\mathbb{P}(y \leq j) = \alpha_j - \beta_1\,x_1 - \cdots - \beta_k\,x_k, \quad j = 1, \ldots, c - 1.$$

- The intercepts are termed `Threshold coefficients`.
- R outputs the individual (non-cumulative) predicted probability for each category.

4.1.7 Example

EXAMPLE 4.1. The California Health Institute Survey (CHIS) is a large-scale survey of the health of Californians. Among the variables measured are gender (M/F), age (in years), marital status ('yes'=married, 'no'=not married), the highest educational degree obtained (<HS/HSgrad/HSgrad+), and the health status (poor/fair/good/excellent). The code below fits the cumulative logit model to the data on 32 respondents.

```
data health_survey;
 length health$ 10.;
   input gender$ age marital$ educ$ health$ @@;
cards;
M 46 yes <HS      good       M 72 yes <HS      poor
M 52 yes HSgrad   excellent  M 50 no  <HS      fair
F 44 no  HSgrad+  poor       F 68 no  HSgrad   fair
F 50 no  HSgrad+  fair       F 93 no  <HS      poor
M 50 yes HSgrad   excellent  M 88 no  HSgrad+  good
M 58 yes HSgrad   excellent  M 52 yes HSgrad   good
F 64 yes HSgrad+  good       F 49 yes HSgrad   good
F 41 yes HSgrad+  excellent  M 32 no  HSgrad+  good
F 88 no  HSgrad   poor       F 36 yes HSgrad+  excellent
M 35 no  HSgrad+  excellent  F 38 no  HSgrad+  fair
M 39 yes HSgrad+  excellent  F 43 no  <HS      good
M 61 yes HSgrad   good       F 41 yes HSgrad+  excellent
F 36 yes <HS      good       F 44 yes HSgrad+  excellent
M 41 no  HSgrad   good       M 55 yes <HS      good
M 57 no  <HS      fair       M 28 yes HSgrad+  excellent
```

```
F 40 yes HSgrad  good        F 97 no  HSgrad  poor
;

proc format;
value $healthfmt 'poor'='1poor' 'fair'='2fair'
  'good'='3good' 'excellent'='4excellent';
run;

proc genmod;
class gender(ref='M') marital(ref='yes')
   educ(ref='HSgrad+');
   model health=gender age marital
      educ/dist=multinomial link=cumlogit;
format health $healthfmt.;
run;
```

Full Log Likelihood   -24.0727
AIC   64.1455
AICC   70.4064
BIC   75.8714

Analysis Of Maximum Likelihood Parameter Estimates

Parameter	Estimate	Pr > ChiSq
Intercept1	-11.7440	<.0001
Intercept2	-9.7021	<.0001
Intercept3	-5.9098	0.0014
gender F	2.0630	0.0281
age	0.0805	0.0145
marital no	3.3593	0.0021
educ <HS	3.2609	0.0093
educ HSgrad	1.1877	0.2420

```
proc genmod;
   model health=/dist=multinomial link=cumlogit;
run;
```

Log Likelihood -41.9644

```
data deviance_test;
 deviance=-2*(-41.9644-(-24.0727));
   pvalue=1-probchi(deviance,5);
run;

proc print;
run;
```

deviance   pvalue
35.7834   0.000001049

In view of the fact that the p-value is tiny, we conclude that the fitted model has a good fit. The fitted model is:

$$\frac{\widehat{\mathbb{P}}(poor)}{1 - \widehat{\mathbb{P}}(poor)} = \exp\Big\{ - 11.7440 + 2.0630 * female$$

$$+0.0805 * age + 3.3593 * not\, married + 3.2609* < HS + 1.1877 * HSgrad\Big\},$$

$$\frac{\widehat{\mathbb{P}}(poor,\, or\, fair)}{1 - \widehat{\mathbb{P}}(poor,\, or\, fair)} = \exp\Big\{ - 9.7021 + 2.0630 * female$$

$$+0.0805 * age + 3.3593 * not\, married + 3.2609* < HS + 1.1877 * HSgrad\Big\},$$

and

$$\frac{\widehat{\mathbb{P}}(poor,\, fair,\, or\, good)}{\widehat{\mathbb{P}}(excellent)} = \exp\Big\{ - 5.9098 + 2.0630 * female$$

$$+0.0805 * age + 3.3593 * not\, married + 3.2609* < HS + 1.1877 * HSgrad\Big\}.$$

The significant predictors at the 5% level are gender, age, marital status, and less than high-school educational level. The corresponding estimated regression coefficients may be interpreted as follows. The estimated odds in favor of worse health rather than better health for females are $\exp\{2.0630\} \cdot 100\% = 786.95\%$ of the estimated odds for males. For a one-year increase in age, these estimated odds increase by $(\exp\{0.0805\} - 1) \cdot 100\% = 8.38\%$. These odds for not married people are estimated to be $\exp\{3.3593\} \cdot 100\% = 2,876.90\%$ of those for married people. For people with less than high school education, these odds are $\exp\{3.2609\} \cdot 100\% = 2,607.30\%$ of those for people with some college education.

Finally, suppose we would like to find the predicted probability for each of the health levels for a married 52-year-old male with a high school diploma. We calculate:

$$\mathbb{P}^0(poor) = \frac{\exp\{-11.7440 + 0.0805 * 52 + 1.1877\}}{1 + \exp\{-11.7440 + 0.0805 * 52 + 1.1877\}} = 0.0017,$$

$$\mathbb{P}^0(poor\, or\, fair) = \frac{\exp\{-9.7021 + 0.0805 * 52 + 1.1877\}}{1 + \exp\{-9.7021 + 0.0805 * 52 + 1.1877\}} = 0.0130,$$

and

$$\mathbb{P}^0(poor,\, fair,\, or\, good) = \frac{\exp\{-5.9098 + 0.0805 * 52 + 1.1877\}}{1 + \exp\{-5.9098 + 0.0805 * 52 + 1.1877\}} = 0.3691.$$

The predicted probabilities for individual categories are: $\mathbb{P}^0(poor) = 0.0017$, $\mathbb{P}^0(fair) = \mathbb{P}^0(poor \text{ or } fair) - \mathbb{P}^0(poor) = 0.0130 - 0.0017 = 0.0113$, $\mathbb{P}^0(good) = \mathbb{P}^0(poor, fair, \text{ or } good) - \mathbb{P}^0(poor \text{ or } fair) = 0.3691 - 0.0130 = 0.3561$, and $\mathbb{P}^0(excellent) = 1 - \mathbb{P}^0(poor, fair, \text{ or } good) = 1 - 0.3691 = 0.6309$.

The following lines of code produce predicted cumulative probabilities in SAS:

```
data prediction;
input gender$ age marital$ educ$;
cards;
M 52 yes HSgrad
;

data health_survey;
 set health_survey prediction;
run;

proc genmod;
class gender(ref='M') marital(ref='yes')
   educ(ref='HSgrad+');
 model health=gender age marital
       educ/dist=multinomial link=cumlogit;
output out=dataout p=pred_prob;
format health $healthfmt.;
run;

proc print data=dataout (firstobs=97 obs=99);
var _level_  pred_prob;
run;
```

LEVEL	pred_prob
1poor	0.00171
2fair	0.01303
3good	0.36934

Thus, the predicted probabilities for individual health categories predicted by SAS are: $\mathbb{P}^0(poor) = 0.00171$, $\mathbb{P}^0(fair) = 0.01303 - 0.00171 = 0.01132$, $\mathbb{P}^0(good) = 0.36934 - 0.01303 = 0.35631$, and $\mathbb{P}^0(excellent) = 1 - 0.36934 = 0.63066$.

The R script and relevant output for this example are below.

```
health.survey.data<-read.csv(file="./Example4.1Data.csv",
header=TRUE, sep=",")
install.packages("ordinal")
library("ordinal")

#specifying reference categories
gender.rel<- relevel(health.survey.data$gender, ref="M")
marital.rel<- relevel(health.survey.data$marital, ref="yes")
educ.rel<- relevel(health.survey.data$educ, ref="HSgrad+")

#fitting cumulative logit model
summary(fitted.model<- clm(health ~ gender.rel + age
+ marital.rel + educ.rel, data=health.survey.data,
link="logit"))
```

AIC
64.15

Coefficients:

	Estimate	Pr(>|z|)
gender.relF	-2.06296	0.02810
age	-0.08052	0.01450
marital.relno	-3.35931	0.00206
educ.rel<HS	-3.26088	0.00926
educ.relHSgrad	-1.18772	0.24200

Threshold coefficients:

	Estimate
1poor|2fair	-11.744
2fair|3good	-9.702
3good|4excellent	-5.910

```
#extracting AICC and BIC for fitted model
p<- 8
n<- 32
print(AICC<- -2*logLik(fitted.model)+2*p*n/(n-p-1))
```

70.40635

```
BIC(fitted.model)
```

75.87137

```
#checking model fit
intercept.only.model<- clm(health ~ 1, data=health.survey.data,
link="logit")
print(deviance<- -2*(logLik(intercept.only.model)
-logLik(fitted.model)))
```

35.78339

```
print(p.value<- pchisq(deviance, df=5, lower.tail=FALSE))
```

1.049431e-06

```
#using fitted model for prediction
print(predict(fitted.model, type="prob",
data.frame(gender.rel="M", age=52, marital.rel="yes",
educ.rel="HSgrad")))
```

```
     1poor       2fair      3good    4excellent
0.001710524  0.01132041  0.3563093  0.6306598
```

□

4.2   Cumulative Probit Model

4.2.1   Model Definition

The *cumulative probit model*[2] for an ordinal response y with values in categories $1, \ldots, c$ and predictors $x_1, \ldots, x_k$ has the form

$$\mathbb{P}(y \le j) = \Phi\big(\alpha_j + \beta_1\, x_1 + \cdots + \beta_k\, x_k\big), \quad j = 1, \ldots, c - 1,$$

where $\Phi(\cdot)$ denotes the cumulative distribution function of a standard normal random variable. Note that like the cumulative logistic regression, the cumulative probit model also belongs to the class of generalized linear regressions, with the probit link function.

---

[2]First proposed in McKelvey, R.D. and W. Zavoina (1975). "A statistical model for the analysis of ordinal level dependent variables". *Journal of Mathematical Sociology*, 4: 103 – 120.

4.2.2 Fitted Model

The fitted model may be written as

$$\Phi^{-1}\big(\widehat{\mathbb{P}}(y \le j)\big) = \widehat{\alpha}_j + \widehat{\beta}_1\,x_1 + \cdots + \widehat{\beta}_k\,x_k, \quad j = 1, \ldots, c - 1, \tag{4.4}$$

or

$$\widehat{\mathbb{P}}(y \le j) = \Phi\big(\widehat{\alpha}_j + \widehat{\beta}_1\,x_1 + \cdots + \widehat{\beta}_k\,x_k\big), \quad j = 1, \ldots, c - 1. \tag{4.5}$$

4.2.3 Interpretation of Estimated Regression Coefficients

In view of (4.4), the estimated cumulative probit regression coefficients are interpreted as described below.

- If a predictor variable x_1 is numeric, then $\widehat{\beta}_1$ gives the change in the estimated z-score for a one-unit increase in x_1, controlling for the other predictors. Indeed,

$$\Phi^{-1}\big(\widehat{\mathbb{P}}(y \le j | x_1 + 1)\big) - \Phi^{-1}\big(\widehat{\mathbb{P}}(y \le j | x_1)\big)$$
$$= \widehat{\beta}_1(x_1 + 1) + \widehat{\beta}_2\,x_2 + \cdots + \widehat{\beta}_k\,x_k - \big(\widehat{\beta}_1\,x_1 + \widehat{\beta}_2\,x_2 + \cdots + \widehat{\beta}_k\,x_k\big) = \widehat{\beta}_1, \quad j = 1, \ldots, c-1.$$

- If a predictor variable x_1 is an indicator variable, then $\widehat{\beta}_1$ represents the difference in the estimated z-scores when $x_1 = 1$ and $x_1 = 0$, provided all the other predictor variables stay unchanged. We write

$$\Phi^{-1}\big(\widehat{\mathbb{P}}(y \le j | x_1 = 1)\big) - \Phi^{-1}\big(\widehat{\mathbb{P}}(y \le j | x_1 = 0)\big)$$
$$= \widehat{\beta}_1 \cdot 1 + \widehat{\beta}_2\,x_2 + \cdots + \widehat{\beta}_k\,x_k - \big(\widehat{\beta}_1 \cdot 0 + \widehat{\beta}_2\,x_2 + \cdots + \widehat{\beta}_k\,x_k\big) = \widehat{\beta}_1, \quad j = 1, \ldots, c-1.$$

4.2.4 Predicted Probabilities

Taking into account (4.5), in the cumulative probit model, predicted cumulative probabilities are computed as

$$\mathbb{P}^0(y \le j) = \Phi\big(\widehat{\alpha}_j + \widehat{\beta}_1\,x_1^0 + \cdots + \widehat{\beta}_k\,x_k^0\big), \quad j = 1, \ldots, c - 1,$$

for predetermined values $x_1^0, \ldots, x_k^0$. Predicted probabilities of exact equality to $j, j = 1, \ldots, c$, are found via the same equations (4.3) as in the cumulative logit model.

4.2.5 SAS Implementation

The procedure `genmod` with `dist=multnomial link=cumprobit` may be used to run the cumulative probit model.

4.2.6 R Implementation

The cumulative probit model may be fit using function `clm()` with the option `link="probit"`.

4.2.7 Example

EXAMPLE 4.2. Consider the data `health_survey` in Example 4.1. We fit the cumulative probit model to the data set, check its goodness-of-fit, and use the fitted model for prediction. The code and important outputs are given below.

```
proc genmod;
class gender(ref='M') marital(ref='yes')
   educ(ref='HSgrad+');
  model health=gender age marital
     educ/dist=multinomial link=cumprobit;
format health $healthfmt.;
run;
```

```
Full Log Likelihood   -24.1213
AIC      64.2426
AICC   70.5035
BIC      75.9685
```

Parameter	Estimate	Pr > ChiSq
Intercept1	-6.7442	<.0001
Intercept2	-5.6721	<.0001
Intercept3	-3.4867	0.0007
gender F	1.1447	0.0257
age	0.0499	0.0062
marital no	1.8788	0.0010
educ <HS	1.8103	0.0057
educ HSgrad	0.6060	0.2942

```
proc genmod data=health_survey;
  model health=/dist=multinomial link=cumprobit;
run;
```

Full Log Likelihood   -41.9644

```
data deviance_test;
 deviance=-2*(-41.9644-(-24.1213));
   pvalue=1-probchi(deviance,5);
run;

proc print;
run;
```

deviance   pvalue
 35.6862   0.000001097

The fitted model has the form:

$$\mathbb{P}(poor) = \Phi\big(-6.7442 + 1.1447 * female + 0.0499 * age$$
$$+1.8788 * not\,married + 1.8103* < HS + 0.6060 * HSgrad\big),$$
$$\mathbb{P}(poor,\,or\,fair) = \Phi\big(-5.6721 + 1.1447 * female + 0.0499 * age$$
$$+1.8788 * not\,married + 1.8103* < HS + 0.6060 * HSgrad\big),$$

and

$$\mathbb{P}(poor,\,fair,\,or\,good) = \Phi\big(-3.4867 + 1.1447 * female + 0.0499 * age$$
$$+1.8788 * not\,married + 1.8103* < HS + 0.6060 * HSgrad\big).$$

We can see that the fitted model has a good fit because the p-value in the deviance test is very small. Similar to the cumulative logistic model, in this case gender, age, marital status, and less than high school educational level are statistically significant predictors. Their estimated beta coefficients may be interpreted as follows. The estimated z-score for women is larger than that for men by 1.1447. As age increases by one year, the estimated z-score increases by 0.0499. People who are not married have estimated z-scores larger by 1.8788 compared to their married counterparts. Finally, estimated z-score for people with less than high school education exceeds that for people who studied in college by 1.8103.

Next, we will find predicted probabilities of each category of health status for a 52-year-old married man who has a high school diploma. To this end, we write:

$$\mathbb{P}^0(poor) = \Phi\big(-6.7442 + 0.0499 * 52 + 0.6060\big) = 0.00020,$$
$$\mathbb{P}^0(poor,\,or\,fair) = \Phi\big(-5.6721 + 0.0499 * 52 + 0.6060\big) = 0.00673,$$

and

$$\mathbb{P}(poor,\ fair,\ or\ good) = \Phi\big(-3.4867 + 0.0499 * 52 + 0.6060\big) = 0.38748.$$

From here, the individual predicted probabilities are $\mathbb{P}^0(poor) = 0.00020$, $\mathbb{P}^0(fair) = 0.00673 - 0.00020 = 0.00653$, $\mathbb{P}^0(good) = 0.38748 - 0.00673 = 0.380746$, and $\mathbb{P}^0(excellent) = 1 - 0.38748 = 0.61252$.

The cumulative probabilities can be requested in SAS by entering the following statements:

```
data prediction;
input gender$ age marital$ educ$;
cards;
M 52 yes HSgrad
;

data health_survey;
 set health_survey prediction;
run;

proc genmod;
class gender(ref='M') marital(ref='yes')
   educ(ref='HSgrad+');
 model health=gender age marital
      educ/dist=multinomial link=cumprobit;
output out=dataout p=pred_prob;
format health $healthfmt.;
run;

proc print data=dataout (firstobs=97 obs=99);
var _level_  pred_prob;
run;
```

LEVEL	pred_prob
1poor	0.00020
2fair	0.00669
3good	0.38658

We can obtain the non-cumulative probabilities by subtraction: $\mathbb{P}^0(poor) = 0.00020$, $\mathbb{P}^0(fair) = 0.00669 - 0.00020 = 0.00649$, $\mathbb{P}^0(good) = 0.38658 - 0.00669 = 0.37989$, and $\mathbb{P}^0(excellent) = 1 - 0.38658 = 0.61342$.

The R script below reproduces the results that were obtained in SAS.

```
health.survey.data<- read.csv(file="./Example4.1Data.csv",
header=TRUE, sep=",")
install.packages("ordinal")
library("ordinal")

#specifying reference categories
gender.rel<- relevel(health.survey.data$gender, ref="M")
marital.rel<- relevel(health.survey.data$marital, ref="yes")
educ.rel<- relevel(health.survey.data$educ, ref="HSgrad+")

#fitting cumulative probit model
summary(fitted.model<- clm(health ~ gender.rel + age
+ marital.rel + educ.rel,
data=health.survey.data, link="probit"))
```

AIC
64.24

Coefficients:

	Estimate	$Pr(>|z|)$
gender.relF	-1.14467	0.025685
age	-0.04986	0.006244
marital.relno	-1.87884	0.000952
educ.rel<HS	-1.81028	0.005683
educ.relHSgrad	-0.60599	0.294219

Threshold coefficients:

	Estimate
1poor|2fair	-6.744
2fair|3good	-5.672
3good|4excellent	-3.487

```
#extracting AICC and BIC for fitted model
p<- 8
n<- 32
print(AICC<- -2*logLik(fitted.model)+2*p*n/(n-p-1))
```

70.50345

```
BIC(fitted.model)
```

75.96847

```
#checking model fit
intercept.only.model<- clm(health ~ 1, data=health.survey.data,
link="probit")
print(deviance<- -2*(logLik(intercept.only.model)
-logLik(fitted.model)))
```

35.68629

```
print(p.value<- pchisq(deviance, df=5, lower.tail=FALSE))
```

1.0974e-06

```
#using fitted model for prediction
print(predict(fitted.model, type="prob",
data.frame(gender.rel="M", age=52, marital.rel="yes",
educ.rel="HSgrad")))
```

1poor	2fair	3good	4excellent
0.0001957589	0.006491104	0.3798914	0.6134217

□

4.3 Cumulative Complementary Log-Log Model

4.3.1 Model Definition

The *cumulative complementary log-log model*[3] for an ordinal response y and predictors $x_1, \ldots, x_k$ is defined by

$$\ln\left[-\ln\left(1 - \mathbb{P}(y \le j)\right)\right] = \alpha_j + \beta_1 x_1 + \cdots + \beta_k x_k, \quad j = 1, \ldots, c-1,$$

or, equivalently,

$$\mathbb{P}(y \le j) = 1 - \exp\left\{-\exp\left\{\alpha_j + \beta_1 x_1 + \cdots + \beta_k x_k\right\}\right\}, \quad j = 1, \ldots, c-1.$$

Similar to cumulative logit and probit models, the cumulative complementary log-log model falls into the class of generalized linear models with the complementary log-log link function.

---

[3]First appears in McCullagh, P. (1980). "Regression models for ordinal data". *Journal of the Royal Statistical Society, Series B*, 42 (2): 109 – 142.

4.3.2 Fitted Model

In the fitted cumulative complementary log-log model, the estimated cumulative probabilities satisfy

$$\widehat{\mathbb{P}}(y \le j) = 1 - \exp\left\{ -\exp\left\{ \widehat{\alpha}_j + \widehat{\beta}_1 x_1 + \cdots + \widehat{\beta}_k x_k \right\} \right\}, \quad j = 1, \ldots, c-1. \quad (4.6)$$

4.3.3 Interpretation of Estimated Regression Coefficients

From (4.6), the estimated regression coefficients are interpreted as:

- If a predictor variable x_1 is numeric, then $\exp\left\{\widehat{\beta}_1\right\}$ gives the estimated *rate ratio* for a unit increase in x_1, given that other predictors remain the same, since for any $j = 1, \ldots, c-1$,

$$1 - \widehat{\mathbb{P}}(y \le j | x_1 + 1) = \exp\left\{ -\exp\left\{ \widehat{\alpha}_j + \widehat{\beta}_1(x_1 + 1) + \widehat{\beta}_2 x_2 + \cdots + \widehat{\beta}_k x_k \right\} \right\}$$

$$= \exp\left\{ -\exp\left\{ \widehat{\alpha}_j + \widehat{\beta}_1 x_1 + \widehat{\beta}_2 x_2 + \cdots + \widehat{\beta}_k x_k \right\} \exp\{\widehat{\beta}_1\} \right\}$$

$$= \left[\exp\left\{ -\exp\left\{ \widehat{\alpha}_j + \widehat{\beta}_1 x_1 + \widehat{\beta}_2 x_2 + \cdots + \widehat{\beta}_k x_k \right\} \right\} \right]^{\exp\{\widehat{\beta}_1\}} = \left[1 - \widehat{\mathbb{P}}(y \le j | x_1) \right]^{\exp\{\widehat{\beta}_1\}}.$$

Thus, when x_1 increases by one unit, the estimated complement cumulative probability $1 - \widehat{\mathbb{P}}(y \le j | x_1 + 1)$ equals to the complement cumulative probability $1 - \widehat{\mathbb{P}}(y \le j | x_1)$ raised to the power $\exp\{\widehat{\beta}_1\}$.

- If a predictor variable x_1 is an indicator variable with two levels 0 and 1, then $\exp\left\{\widehat{\beta}_1\right\}$ gives the estimated *rate ratio* for the levels $x_1 = 1$ and $x_1 = 0$, provided the other predictors stay intact, which can be seen by writing

$$1 - \widehat{\mathbb{P}}(y \le j | x_1 = 1) = \exp\left\{ -\exp\left\{ \widehat{\alpha}_j + \widehat{\beta}_1 \cdot 1 + \widehat{\beta}_2 x_2 + \cdots + \widehat{\beta}_k x_k \right\} \right\}$$

$$= \left[\exp\left\{ -\exp\left\{ \widehat{\alpha}_j + \widehat{\beta}_1 \cdot 0 + \widehat{\beta}_2 x_2 + \cdots + \widehat{\beta}_k x_k \right\} \right\} \right]^{\exp\{\widehat{\beta}_1\}}$$

$$= \left[1 - \widehat{\mathbb{P}}(y \le j | x_1 = 0) \right]^{\exp\{\widehat{\beta}_1\}}, \quad j = 1, \ldots, c-1.$$

4.3.4 Predicted Probabilities

Using (4.6), we can write the equations for the predicted cumulative probabilities for some defined values of predictors $x_1^0, \ldots, x_k^0$ as

$$\mathbb{P}^0(y \le j) = 1 - \exp\left\{ -\exp\left\{ \widehat{\alpha}_j + \widehat{\beta}_1 x_1^0 + \cdots + \widehat{\beta}_k x_k^0 \right\} \right\}, \quad j = 1, \ldots, c-1.$$

The individual probabilities may be obtained via the expressions in (4.3).

4.3.5   SAS Implementation

To fit the cumulative complementary log-log model, use `proc genmod` with the option `dist=multimonial link=cumcll`.

4.3.6   R Implementation

The `clm()` function with the option `link="cloglog"` would fit the cumulative complement log-log model.

4.3.7   Example

EXAMPLE 4.3. We will fit the cumulative complementary log-log model to the data in Example 4.1. Referring to the data set `health_survey`, we run the following statements. The relevant output is also presented.

```
proc genmod;
class gender(ref='M') marital(ref='yes')
   educ(ref='HSgrad+');
  model health=gender age marital
     educ/dist=multinomial link=cumcll;
format health $healthfmt.;
run;
```

```
Full Log Likelihood   -23.6572
AIC    63.3144
AICC   69.5753
BIC    75.0403
```

Analysis Of Maximum Likelihood Parameter Estimates

Parameter	Estimate	Pr > ChiSq
Intercept1	-9.2186	<.0001
Intercept2	-7.7333	<.0001
Intercept3	-4.7078	0.0003
gender F	1.2741	0.0239
age	0.0624	0.0046
marital no	2.2980	0.0029

```
educ <HS       2.2481        0.0034
educ HSgrad    0.9700        0.1353
```

```
proc genmod;
  model health=/dist=multinomial link=cumcll;
run;
```

```
Log Likelihood   -41.9644
```

```
data deviance_test;
 deviance=-2*(-41.9644-(-23.6572));
   pvalue=1-probchi(deviance,5);
run;
```

```
proc print;
run;
```

```
deviance          pvalue
 36.6144   0.000000716
```

The fitted model takes the form: $\widehat{\mathbb{P}}(poor) = 1 - \exp\left\{-\exp\left\{-9.2186 + 1.2741 * female + 0.0624 * age + 2.2980 * not\, married + 2.2481 * < HS + 0.9700 * HSgrad\right\}\right\}$, $\widehat{\mathbb{P}}(poor, \text{ or } fair) = 1 - \exp\left\{-\exp\left\{-7.7333 + 1.2741 * female + 0.0624 * age + 2.2980 * not\, married + 2.2481 * < HS + 0.9700 * HSgrad\right\}\right\}$, and $\widehat{\mathbb{P}}(poor, fair, \text{ or } good) = 1 - \exp\left\{-\exp\left\{-4.7078 + 1.2741 * female + 0.0624 * age + 2.2980 * not\, married + 2.2481 * < HS + 0.9700 * HSgrad\right\}\right\}$.

Like in the cumulative logit and probit models, in this model gender, age, marital status, and less than high school educational level are significant at the 5% level. The estimated probability of better health for women is that for men raised to the power $\exp\left\{1.2741\right\} = 3.5755$. For a one-year increase in age, this estimated probability is the old one raised to the power $\exp\{0.0624\} = 1.0644$. For a not married person, the estimated probability is that for a married one raised to the power $\exp\{2.2980\} = 9.9543$. For those with less than high school education, the estimated probability is the one for individuals with some college raised to the power $\exp\{2.2481\} = 9.4697$.

For a 52-year-old men who is married and has graduated from a high school, the predicted cumulative probabilities can be found as

$$\widehat{\mathbb{P}}(poor) = 1 - \exp\left\{-\exp\left\{-9.2186 + 0.0624 * 52 + 0.9700\right\}\right\} = 0.00669,$$

$$\widehat{\mathbb{P}}(poor, \text{ or } fair) = 1 - \exp\left\{-\exp\left\{-7.7333 + 0.0624*52 + 0.9700\right\}\right\} = 0.02921,$$

and

$$\widehat{\mathbb{P}}(poor, fair, \text{ or } good) = 1 - \exp\left\{-\exp\left\{-4.7078 + 0.0624*52 + 0.9700\right\}\right\} = 0.45708.$$

The predicted probabilities for individual categories are computed as $\widehat{\mathbb{P}}(poor) = 0.00669$, $\widehat{\mathbb{P}}(fair) = 0.02921 - 0.00669 = 0.02252$, $\widehat{\mathbb{P}}(good) = 0.45708 - 0.02921 = 0.42787$, and $\widehat{\mathbb{P}}(excellent) = 1 - 0.45708 = 0.54292$. The same predictions for cumulative probabilities are made by SAS via the following statements:

```
data prediction;
input gender$ age marital$ educ$;
cards;
M 52 yes HSgrad
;

data health_survey;
 set health_survey prediction;
run;

proc genmod;
class gender(ref='M') marital(ref='yes')
   educ(ref='HSgrad+');
 model health=gender age marital
      educ/dist=multinomial link=cumcll;
output out=dataout p=pred_prob;
format health $healthfmt.;
run;

proc print data=dataout (firstobs=97 obs=99);
var _level_  pred_prob;
run;
```

LEVEL	pred_prob
1poor	0.00669
2fair	0.02921
3good	0.45708

The R script and the output for this example follow.

```
health.survey.data<- read.csv(file="./Example4.1Data.csv",
header=TRUE, sep=",")
install.packages("ordinal")
library("ordinal")

#specifying reference categories
gender.rel<- relevel(health.survey.data$gender, ref="M")
marital.rel<- relevel(health.survey.data$marital, ref="yes")
educ.rel<- relevel(health.survey.data$educ, ref="HSgrad+")

#fitting cumulative complementary log-log model
summary(fitted.model<- clm(health ~ gender.rel + age
+ marital.rel + educ.rel, data=health.survey.data,
link="cloglog"))
```

AIC
63.31

Coefficients:

	Estimate	Pr(>|z|)
gender.relF	-1.27412	0.02392
age	-0.06240	0.00458
marital.relno	-2.29800	0.00288
educ.rel<HS	-2.24809	0.00337
educ.relHSgrad	-0.97005	0.13527

Threshold coefficients:

	Estimate
1poor|2fair	-9.219
2fair|3good	-7.733
3good|4excellent	-4.708

```
#extracting AICC and BIC for the fitted probit model
p<- 8
n<- 32
print(AICC<- -2*logLik(fitted.model)+2*p*n/(n-p-1))
```

69.57525

```
BIC(fitted.model)
```

75.04027

```
#checking model fit
intercept.only.model<- clm(health ~ 1, data=health.survey.data,
link="cloglog")
print(deviance<- -2*(logLik(intercept.only.model)
-logLik(fitted.model)))
```

36.6145

```
print(p.value<- pchisq(deviance, df=5, lower.tail=FALSE))
```

7.155456e-07

```
#using fitted model for prediction
print(predict(fitted.model, type="prob",
data.frame(gender.rel="M", age=52, marital.rel="yes",
educ.rel="HSgrad")))
```

```
         1poor       2fair      3good    4excellent
1 0.006689841  0.02251738  0.4278729   0.5429199
```

The last thing we do in this example is to select the best-fitted model among the cumulative logit, probit, and complementary log-log models. Here we summarize the values for the AIC, AICC, and BIC criteria that were given in the outputs for these models.

	cumlogit	cumprobit	cumcloglog
AIC	64.15	64.24	63.31
AICC	70.4064	70.5035	69.5753
BIC	75.8714	75.9685	75.0403

The values are the smallest for the cumulative complementary log-log model, thus it fits the data the best and should be preferred over the other two models. □

4.4 Generalized Logit Model for Nominal Response

4.4.1 Model Definition

Suppose the response y is a nominal variable with categories $j = 1, \ldots, c$. Let the category c be the reference category. The *generalized logit functions* pair each category with the reference category, that is,

$$\text{logit } \mathbb{P}(y = j) = \ln \frac{\mathbb{P}(y = j)}{\mathbb{P}(y = c)}, \quad j = 1, \ldots, c - 1.$$

If, for example, $c = 3$, then

$$\text{logit } \mathbb{P}(y = 1) = \ln \frac{\mathbb{P}(y = 1)}{\mathbb{P}(y = 3)}, \quad \text{and } \text{logit } \mathbb{P}(y = 2) = \ln \frac{\mathbb{P}(y = 2)}{\mathbb{P}(y = 3)}.$$

The *generalized logit model*[4] for a nominal response y and predictors $x_1, \dots, x_k$ is given by

$$\text{logit } \mathbb{P}(y = j) = \alpha_j + \beta_{j1} x_1 + \cdots + \beta_{jk} x_k, \quad j = 1, \dots, c - 1.$$

Note that in this model all regression coefficients are different. There are $c - 1$ intercepts and $(c - 1) k$ beta coefficients. This model doesn't belong to the class of generalized linear regressions.

4.4.2   Fitted Model

Once the regression coefficients are estimated, the fitted model may be written as

$$\ln \frac{\widehat{\mathbb{P}}(y = j)}{\widehat{\mathbb{P}}(y = c)} = \widehat{\alpha}_j + \widehat{\beta}_{j1} x_1 + \cdots + \widehat{\beta}_{jk} x_k, \quad j = 1, \dots, c - 1,$$

or, equivalently,

$$\frac{\widehat{\mathbb{P}}(y = j)}{\widehat{\mathbb{P}}(y = c)} = \exp \left\{ \widehat{\alpha}_j + \widehat{\beta}_{j1} x_1 + \cdots + \widehat{\beta}_{jk} x_k \right\} \quad j = 1, \dots, c - 1. \qquad (4.7)$$

4.4.3   Interpretation of Estimated Regression Coefficients

The estimated regression coefficients are interpreted as follows.

• If a predictor variable x_1 is numeric, then the quantity $\left(\exp\{\widehat{\beta}_{j1}\} - 1 \right) \cdot 100\%$ represents the percent change in estimated odds of the event $\{y = j\}$ as opposed to $\{y = c\}$ for a unit increase in x_1, controlling for all the other predictors. Indeed,

$$\left(\frac{\widehat{\mathbb{P}}(y = j | x_1 + 1) / \widehat{\mathbb{P}}(y = c | x_1 + 1) - \widehat{\mathbb{P}}(y = j | x_1) / \widehat{\mathbb{P}}(y = c | x_1)}{\widehat{\mathbb{P}}(y = j | x_1) / \widehat{\mathbb{P}}(y = c | x_1)} \right) \cdot 100\%$$

---

[4]A common reference to the primary source is Agresti, A. (1990). "*Categorical Data Analysis*," Wiley, New York.

$$= \left(\frac{\exp\{\widehat{\alpha}_j + \widehat{\beta}_{j1}(x_1 + 1) + \widehat{\beta}_{j2}\, x_2 + \cdots + \widehat{\beta}_{jk}\, x_k\}}{\exp\{\widehat{\alpha}_j + \widehat{\beta}_{j1}\, x_1 + \widehat{\beta}_{j2}\, x_2 + \cdots + \widehat{\beta}_{jk}\, x_k\}} - 1 \right) \cdot 100\%$$

$$= \left(\exp\{\widehat{\beta}_{j1}\} - 1 \right) \cdot 100\%.$$

- If a predictor x_1 is a 0-1 variable, then $\exp\{\widehat{\beta}_{j1}\} \cdot 100\%$ gives the ratio of the estimated odds of $\{y = j\}$ versus $\{y = c\}$ for the levels $x_1 = 1$ and $x_1 = 0$, provided the other predictors stay unchanged. We write

$$\left(\frac{\widehat{\mathbb{P}}(y = j | x_1 = 1) / \widehat{\mathbb{P}}(y = c | x_1 = 1)}{\widehat{\mathbb{P}}(y = j | x_1 = 0) / \widehat{\mathbb{P}}(y = c | x_1 = 0)} \right) \cdot 100\%$$

$$= \frac{\exp\{\widehat{\alpha}_j + \widehat{\beta}_{j1} \cdot 1 + \widehat{\beta}_{j2}\, x_2 + \cdots + \widehat{\beta}_{jk}\, x_k\}}{\exp\{\widehat{\alpha}_j + \widehat{\beta}_{j1} \cdot 0 + \widehat{\beta}_{j2}\, x_2 + \cdots + \widehat{\beta}_{jk}\, x_k\}} \cdot 100\% = \exp\{\widehat{\beta}_{j1}\} \cdot 100\%.$$

4.4.4 Predicted Probabilities

In view of (4.7), for a fixed set of predictors $x_1^0, \ldots, x_k^0$, the predicted probability that $y = j$, where $j = 1, \ldots, c - 1$, is

$$\mathbb{P}^0(y = j) = \mathbb{P}^0(y = c) \, \exp \left\{ \widehat{\alpha}_j + \widehat{\beta}_{j1}\, x_1^0 + \cdots + \widehat{\beta}_{jk}\, x_k^0 \right\}, \qquad (4.8)$$

and, since the probabilities must add up to one,

$$\mathbb{P}^0(y = c) = \left(1 + \sum_{j=1}^{c-1} \exp \left\{ \widehat{\alpha}_j + \widehat{\beta}_{j1}\, x_1^0 + \cdots + \widehat{\beta}_{jk}\, x_k^0 \right\} \right)^{-1}. \qquad (4.9)$$

4.4.5 SAS Implementation

The generalized logit model may be fitted by using the `logistic` procedure with the option `link=glogit`. The syntax is:

```
proc logistic data=data_name;
  class catpredictor1_name (ref='level_name') catpredictor2_name
        (ref='level_name') .../param=ref;
model response_name = <list of predictors>/link=glogit;
    output out=outdata_name p=predicted_response_name;
run;
```

- The option `param=ref` must be included in the `class` statement in order for proper indicator functions to be created. If this option is omitted, then SAS creates variables that assume values 1 and -1 instead of 1 and 0.

94

4.4.6　R Implementation

The function `multinom()` from the package `nnet` (stands for "neural networks") may be used to fit the generalized logit model. The syntax is

`summary(`*fitted.model.name* `<- multinom(`*response.name* `~` *x1.name* `+` $\cdots$ `+` *xk.name*`, data=`*data.name*`))`

4.4.7　Example

EXAMPLE 4.4. In a dental clinic, an oral hygienist surveys new patients in order to find out what factors influence their choice of toothpaste. For each patient, she records gender, age, the number of problematic teeth (teeth with fillings, root canals, or extracted), and the type of toothpaste used (tartar control, cavity protection, or for sensitive teeth). The following SAS code fits a generalized logit model for the choice of toothpaste.

```
data oral_hygene;
 length choice$ 9.;
input ID gender$ age nteeth choice$ @@;
cards;
1   F 48 7   sensitive    2   M 30 5 cavity
3   M 34 6   tartar       4   M 50 8 sensitive
5   M 68 4   tartar       6   F 22 1 cavity
7   M 53 8   sensitive    8   F 38 2 cavity
9   M 36 7   sensitive    10 F 25 1 cavity
11 F 32 7   sensitive    12 F 54 2 tartar
13 M 32 8   sensitive    14 M 26 3 cavity
15 F 35 2   cavity       16 F 33 8 cavity
17 F 52 4   tartar       18 F 43 9 sensitive
19 M 58 2   tartar       20 F 43 3 cavity
21 F 60 6   tartar       22 M 28 3 tartar
23 M 70 10 sensitive    24 M 41 2 tartar
25 F 43 5   cavity       26 M 18 1 cavity
27 M 66 12 sensitive    28 M 34 2 sensitive
;

proc logistic;
 class gender(ref='M')/param=ref;
   model choice=gender age nteeth/link=glogit;
run;
```

The output is

Model Fit Statistics

Criterion	Intercept Only	Intercept and Covariates
-2 Log L	61.229	25.370

Analysis of Maximum Likelihood Estimates

Parameter		choice	Estimate	Pr > ChiSq
Intercept		cavity	9.7152	0.0310
Intercept		sensitive	-1.3987	0.6361
gender	F	cavity	5.0207	0.0369
gender	F	sensitive	0.4061	0.8106
age		cavity	-0.3176	0.0143
age		sensitive	-0.0903	0.2323
nteeth		cavity	0.00143	0.9976
nteeth		sensitive	0.9806	0.0164

To conduct the deviance test, we submit the following statements:

```
data deviance_test;
 deviance=61.229-25.370;
   pvalue=1-probchi(deviance,3);
run;

proc print;
run;
```

deviance	pvalue
35.859	8.0204E-8

The p-value is extremely small compared to 0.05. Thus, the model has a good fit. The fitted model is

$$\frac{\widehat{\mathbb{P}}(cavity)}{\widehat{\mathbb{P}}(tartar)} = \exp\left\{9.7152 + 5.0207 * female - 0.3176 * age + 0.00143 * nteeth\right\},$$

and

$$\frac{\widehat{\mathbb{P}}(sensitive)}{\widehat{\mathbb{P}}(tartar)} = \exp\left\{-1.3987 + 0.4061 * female - 0.0903 * age + 0.9806 * nteeth\right\},$$

Gender and age are significant predictors of odds in favor of cavity protection versus tartar control toothpaste, since both p-values are less than 0.05. The estimated odds in favor of cavity protection versus tartar control toothpaste for women is $\exp\{5.0207\} \cdot 100\% = 15,151.73\%$ of those for men. Also, as age increases by one year, the estimated odds change by $(\exp\{-0.3176\} - 1) \cdot 100\% = -27.21\%$, that is, decrease by 27.21%.

The number of problematic teeth is the only significant predictor of odds in favor of toothpaste for sensitive teeth versus tartar control toothpaste, since its p-value is less than 0.05. As the number of problematic teeth increases by one, the estimated odds grow by $(\exp\{0.9806\} - 1) \cdot 100\% = 166.61\%$.

To estimate the probabilities of choosing each type of toothpaste by a 49-year-old male patient with seven problematic teeth, we compute $\mathbb{P}^0(cavity) = \exp\{9.7152 - 0.3176 * 49 + 0.00143 * 7\}(1 + \exp\{9.7152 - 0.3176 * 49 + 0.00143 * 7\} + \exp\{-1.3987 - 0.0903 * 49 + 0.9806 * 7\})^{-1} = 0.00076$, $\mathbb{P}^0(sensitive) = \exp\{-1.3987 - 0.0903 * 49 + 0.9806 * 7\}(1 + \exp\{9.7152 - 0.3176 * 49 + 0.00143 * 7\} + \exp\{-1.3987 - 0.0903 * 49 + 0.9806 * 7\})^{-1} = 0.73844$, $\mathbb{P}^0(tartar) = (1 + \exp\{9.7152 - 0.3176 * 49 + 0.00143 * 7\} + \exp\{-1.3987 - 0.0903 * 49 + 0.9806 * 7\})^{-1} = 0.26080$.

SAS outputs similar predicted values, which can be checked by running the following lines of code.

```
data prediction;
input gender$ age nteeth;
cards;
M 49 7
;

data oral_hygene;
 set oral_hygene prediction;
run;

proc logistic;
class gender(ref='M')/param=ref;
 model choice=gender age nteeth/link=glogit;
  output out=outdata p=pred_prob;
run;

proc print data=outdata (firstobs=85 obs=87);
var _level_  pred_prob;
```

```
run;
```

The output is:

gender	age	nteeth	_LEVEL_	pred_prob
M	49	7	cavity	0.00076
M	49	7	sensitive	0.73839
M	49	7	tartar	0.26085

The statements in R that produce the same results as in SAS and the corresponding relevant outputs are located below.

```
oralhygene.data<-read.csv(file="./Example4.4Data.csv",
header=TRUE, sep=",")
install.packages("nnet")
library(nnet)

#specifying reference categories
gender.rel<- relevel(oralhygene.data$gender, ref="M")
choice.rel<- relevel(oralhygene.data$choice, ref="tartar")

#fitting generalized logits model
summary(fitted.model<- multinom(choice.rel ~ gender.rel
 + age + nteeth, data = oralhygene.data))
```

Coefficients:

	(Intercept)	gender.relF	age	nteeth
cavity	9.715717	5.0209204	-0.3175656	0.001382412
sensitive	-1.398798	0.4061775	-0.0903070	0.980622823

```
#checking model fit
summary(intercept.model<- multinom(choice.rel ~ 1,
data=oralhygene.data))
print(deviance<- deviance(intercept.model)
-deviance(fitted.model))
```

35.85856

```
print(p.value <- pchisq(deviance, df=3,lower.tail = FALSE))
```

8.022154e-08

```
#using fitted model for prediction
print(predict(fitted.model, type="probs",
data.frame(gender.rel="M", age=49, nteeth=7)))
```

tartar	cavity	sensitive
0.2608508757	0.0007623345	0.7383867899

□

Exercises for Chapter 4

EXERCISE 4.1. Grade point average (GPA) and graduate management apti-tude test (GMAT) scores are used by admission office of a business school to decide which applicant should be admitted to the school's graduate program. The data below are GPA and GMAT scores for 42 recent applicants who have been categorized as admitted, borderline, or not admitted.

GPA	GMAT	Status	GPA	GMAT	Status
2.96	596	admit	2.8	444	border
3.14	473	admit	3.13	416	border
3.22	482	admit	2.89	431	border
3.29	527	admit	3.01	471	border
3.69	505	admit	2.91	446	border
2.46	693	admit	2.75	546	border
3.03	626	admit	2.73	467	border
3.19	663	admit	3.12	463	border
3.63	447	admit	3.08	440	notadmit
3.59	588	admit	3.01	453	notadmit
3.3	563	admit	3.03	414	notadmit
3.78	591	admit	3.04	446	notadmit
3.44	692	admit	2.89	485	notadmit
3.48	528	admit	2.79	490	notadmit
3.47	552	admit	2.54	446	notadmit
3.35	520	admit	2.43	425	notadmit
2.89	543	admit	2.2	474	notadmit
2.28	523	admit	3.36	531	notadmit
3.21	530	admit	2.57	542	notadmit
3.58	564	admit	2.36	482	notadmit
3.33	565	admit	3.66	420	notadmit

(a) Run the cumulative logit model and specify the fitted model. Discuss the model fit. What predictors are significant at the 5% level? Interpret the esti-mated significant regression coefficients. Predict the probabilities of each ad-mission status for a person whose GPA is 3.1 and GMAT score is 550.

(b) Redo part (a), fitting the cumulative probit model.

(c) Redo part (a), fitting the cumulative complementary log-log model.

(d) Which of the models obtained in parts (a)-(c) has the best fit?

EXERCISE 4.2. A satellite television provider is focusing on improving customer service. The company surveys subscribers who contact the call center and record how long the callers have been subscribed with the company (in months), whether they receive their monthly programming magazine (yes/no), whether the issue they called about was resolved (yes/no), and overall satisfaction with the customer service, measured on a five-point Likert scale (1=very dissatisfied, 2=dissatisfied, 3=neutral, 4=satisfied, 5=very satisfied). The data for 36 callers are recorded below.

Subscr	Magzn	Resolved	Satisf	Subscr	Magzn	Resolved	Satisf
5	yes	no	5	2	no	yes	4
49	yes	no	5	11	no	no	2
56	no	no	3	98	yes	yes	5
13	yes	yes	5	11	no	yes	5
27	no	yes	4	46	no	no	4
41	yes	yes	5	7	no	no	3
2	yes	yes	5	7	no	yes	5
64	yes	yes	4	9	yes	yes	5
88	yes	yes	4	17	no	no	2
43	yes	yes	4	8	no	yes	2
94	yes	no	4	9	no	yes	1
8	no	no	1	95	no	no	4
9	yes	no	2	60	no	yes	3
68	yes	no	4	80	no	yes	4
5	no	yes	2	2	yes	no	3
108	no	yes	3	33	yes	yes	4
21	yes	yes	4	5	yes	no	3
25	yes	no	3	7	no	no	1

(a) Regress the satisfaction score on the other variables via the cumulative logit model. How good is the model fit? Which regression coefficients are significant at $\alpha = 0.05$? State the fitted model explicitly and interpret the estimated significant beta coefficients. Predict probabilities of each of the five levels of the satisfaction score for a caller who had been subscribed for 3 months, doesn't receive the magazine, and whose issue was resolved over the phone.

(b) Redo part (a), running the cumulative probit model.

(c) Redo part (a), running the cumulative complementary log-log model.

(d) Discuss the relative fit of the models obtained in parts (a)-(c).

EXERCISE 4.3. A large office supply corporation has collected data on purchasing habits of their customers (which are companies). The data for a sample of 34 companies that purchased products during one-month period are made available. The record is kept on how long the company has been in business (<1 year, 1-5 years, or 5+ years), whether it is a first-time buyer (yes/no), what type of supply is bought (stationary, electronics, or furniture), and the amount of money spent (in U.S. dollars). The data are:

In business	1st time	Type	Amount	In business	1st time	Type	Amount
< 1 year	yes	stationary	$5,690	1-5 years	no	electronics	$24,336
1-5 years	yes	stationary	$4,454	5+ years	yes	stationary	$452
5+ years	yes	electronics	$20,489	< 1 year	yes	stationary	$3,600
5+ years	no	stationary	$3,115	5+ years	yes	furniture	$2,450
< 1 year	no	electronics	$44,885	< 1 year	no	electronics	$12,230
< 1 year	no	electronics	$28,182	5+ years	yes	stationary	$2,451
< 1 year	no	furniture	$40,982	1-5 years	no	stationary	$1,110
< 1 year	no	stationary	$10,160	< 1 year	yes	electronics	$69,280
1-5 years	no	furniture	$51,363	< 1 year	yes	furniture	$119,613
5+ years	yes	electronics	$29,448	< 1 year	no	electronics	$21,770
5+ years	no	stationary	$2,093	< 1 year	yes	electronics	$64,160
< 1 year	no	furniture	$127,133	< 1 year	no	furniture	$78,900
1-5 years	yes	furniture	$21,593	< 1 year	no	electronics	$75,095
< 1 year	no	furniture	$220,909	5+ years	no	furniture	$7,450
1-5 years	no	electronics	$17,000	5+ years	no	furniture	$5,200
1-5 years	yes	electronics	$22,812	< 1 year	no	furniture	$32,099
1-5 years	yes	electronics	$13,090	5+ years	no	electronics	$1,997

(a) Categorize the amount spent into the three categories '<$10,000 ', '$10,000-$30,000', and '$30,000+'. Fit a cumulative logit model. Write down the fitted model, discuss its fit, and interpret estimated significant coefficients. Predict probabilities of each expenditure bracket for a company that has been in business for 4 years, and buys electronics from the supply corporation on the regular basis.

(b) Fit a cumulative probit model to the data, and answer the questions in part (a).

(c) Repeat part (a) with a cumulative complementary log-log model.

(d) Does any of the three models have a strikingly better fit?

EXERCISE 4.4. In aviation, the weather forecast often plays a decisive role. A data set for 30 large airports around the country was obtained. The independent variables are airport elevation (in feet above the sea level), its proximity to a large body of water (whether within 20 miles of lake, sea, or ocean), wind direction (in degrees, clockwise from north), and wind speed (in knots=1.15mph).

The dependent variable is the outcome of the weather forecast: correct prediction, false alarm (when the actual conditions were better than predicted), or failure to detect (when the actual conditions were worse than predicted). The measurements are as follows:

Elev	Water	Wdir	Wspeed	Outcome	Elev	Water	Wdir	Wspeed	Outcome
146	yes	270	2	FA	1026	no	290	1	C
841	no	360	13	FA	17	yes	180	2	C
672	yes	360	4	FA	20	yes	270	6	C
312	no	250	5	FA	15	yes	0	3	C
126	yes	170	8	FA	1135	no	20	13	C
607	no	360	8	FA	21	yes	30	8	C
748	no	270	15	FA	98	no	140	8	C
620	yes	290	5	FA	36	yes	10	3	C
5431	no	200	2	FD	8	yes	270	10	C
2181	yes	310	8	FD	26	yes	0	3	C
645	yes	170	7	FD	13	yes	170	9	C
433	no	270	6	FD	9	yes	270	6	C
360	no	140	15	FD	18	yes	200	12	C
4227	yes	200	2	FD	96	no	200	8	C
14	yes	150	7	C	60	yes	240	9	C

(a) Assuming that the outcome is measured on the nominal scale, run the generalized logit model. Use correct prediction as the reference category. Write down the fitted model explicitly.

(b) How good is the model fit? Which variables are significant predictors at the 10% level of significance?

(c) Give interpretation of the estimated significant coefficients.

(d) Find predicted probabilities of each outcome of weather forecast for an airport that is located at 2,000 feet above the sea level, away from a large body of water, in the presence of wind at 5 knots blowing from the east.

EXERCISE 4.5. A group of 25 school-age patients in an orthopedic clinic are studied and their age, gender and ankle condition (sprained, torn ligament, or broken) are recorded. The data are

Age	Gender	Condition	Age	Gender	Condition
7	female	sprained	10	female	sprained
9	male	torn	9	female	torn
11	male	broken	8	male	sprained
12	male	broken	8	female	sprained
8	male	torn	7	female	torn
8	female	torn	15	male	broken
9	female	broken	17	male	broken
13	male	broken	18	male	broken
13	male	torn	18	female	sprained
15	female	sprained	18	female	torn
16	female	sprained	16	female	torn
11	male	torn	12	male	broken
12	male	broken			

(a) Regress the ankle condition on age and gender by running the generalized logit regression model for the nominal response. Use "sprained" as the reference category.

(b) Write down the estimated model. Discuss its goodness-of-fit.

(c) Interpret the estimates of the regression coefficients that significantly differ from zero.

(d) What are the predicted probabilities of each type of ankle injury for a 9-year-old girl?

EXERCISE 4.6. A sample of 40 female users who were matched with male candidates was obtained from a dating website. The following variables were computed: communication status (0 if neither sent messages, 1 if the user sent a message, 2 if the candidate sent a message, and 3 if they exchanged messages), age difference between the user and candidate (in years), their height difference (in inches), and an indicator of same drinking preferences (1=same, 0=differen.). The data are presented in the table below.

Status	Agediff	Heightdiff	Drinking	Status	Agediff	Heightdiff	Drinking
3	-3	-1	0	2	0	-7	1
3	3	-2	1	2	4	-3	0
3	2	-3	1	1	8	-7	1
3	0	1	1	1	1	0	1
3	-5	0	1	1	11	0	0
3	-6	-6	1	1	-4	-7	0
3	2	-5	1	1	7	-6	1
3	0	-4	1	1	14	-6	1
3	4	-7	1	1	-1	-8	0
3	-1	-8	1	1	-5	-4	0
3	-5	1	1	1	-1	-7	0
3	-2	2	1	1	-3	-8	1
3	-6	-4	1	1	8	-4	1
3	-7	-6	0	1	4	-5	1
2	-5	-1	0	0	4	-8	1
2	-18	0	1	0	-6	3	0
2	-8	3	0	0	-4	3	0
2	4	0	1	0	8	-2	0
2	-4	2	1	0	-5	3	1
2	1	-8	1	0	-6	3	0

(a) Regress the communication status on the other variables. Treat it as a nominal variable. Use the zero level as reference. Write down the fitted model.

(b) Evaluate goodness-of-fit of the model. What predictors are significant at the 5% level of significance?

(c) Give interpretation of the estimated significant beta coefficients.

(d) Describe the situation when the user is least likely to contact her candidate. When is the candidate least likely to message the user? When are both least likely to exchange messages?

Chapter 5

Regression Models for Count Response

Suppose the response variable y assumes values 0, 1, 2, ..., but large values are very unlikely. In this chapter we consider four regressions that may be applied to model this response: Poisson regression, zero-truncated Poisson (if y is strictly positive), zero-inflated Poisson (if zero is an allowed value for y, and, moreover, there are too many zeros observed than can be accounted for by the Poisson distribution), and hurdle Poisson (where zeros are modeled separately from the positive values of y).

5.1 Poisson Regression Model

5.1.1 Model Definition

A variable that assumes only non-negative integer values (0, 1, 2, ...) is called a *count variable*. When the response is a count variable which follows a Poisson distribution, the data may be modeled using a Poisson regression. The *Poisson regression model*[1] specifies that given predictors $x_1, \ldots, x_k$, the response variable y follows a Poisson distribution with the probability mass function

$$\mathbb{P}(Y = y) = \frac{\lambda^y \exp\{-\lambda\}}{y!}, \quad y = 0, 1, 2, \ldots,$$

---

[1] The first application appears in Cochran, W.G. (1940). "The analysis of variance when experimental errors follow the Poisson or binomial law". *Annals of Mathematical Statistics*, 11(3): 335 – 347.

where the rate

$$\lambda = \mathbb{E}(y) = \exp\left\{\beta_0 + \beta_1 x_1 + \cdots + \beta_k x_k\right\}. \tag{5.1}$$

This model belongs to the class of generalized linear models with the log link function. It can be shown (do it!) that the above function can be written in the form (1.3) with $\theta = \ln\lambda$ and $\phi = 1$.

5.1.2 Fitted Model

By (5.1), in a fitted Poisson regression model, the estimated mean response has the form

$$\widehat{\lambda} = \exp\left\{\widehat{\beta}_0 + \widehat{\beta}_1 x_1 + \cdots + \widehat{\beta}_k x_k\right\}. \tag{5.2}$$

5.1.3 Interpretation of Estimated Regression Coefficients

From (5.2), in the Poisson regression model, the estimates of the regression coefficients are interpreted as follows.

- If a predictor variable x_1 is numeric, then the exponentiated estimate of the respective regression coefficient $\exp\{\widehat{\beta}_1\}$ represents an estimated *rate ratio* corresponding to a unit increase in the predictor. Indeed,

$$\frac{\widehat{\lambda}(x_1 + 1, x_2, \ldots, x_k)}{\widehat{\lambda}(x_1, x_2, \ldots, x_k)} = \frac{\exp\left\{\widehat{\beta}_0 + \widehat{\beta}_1\,(x_1 + 1) + \widehat{\beta}_2\,x_2 + \cdots + \widehat{\beta}_k\,x_k\right\}}{\exp\left\{\widehat{\beta}_0 + \widehat{\beta}_1\,x_1 + \cdots + \widehat{\beta}_k\,x_k\right\}} = \exp\left\{\widehat{\beta}_1\right\}.$$

Equivalently, $(\exp\{\widehat{\beta}_1\} - 1) \cdot 100\%$ may be interpreted as the estimated percent change in rate when x_1 increases by one unit, while all the other predictors are held fixed.

- If a predictor variable x_1 is an indicator variable, then the exponentiated estimated coefficient $\exp\{\widehat{\beta}_1\}$ represents the ratio of the estimated rates when $x_1 = 1$ and when $x_1 = 0$. To see that, we write

$$\frac{\widehat{\lambda}(x_1 = 1, x_2, \ldots, x_k)}{\widehat{\lambda}(x_1 = 0, x_2, \ldots, x_k)} = \frac{\exp\left\{\widehat{\beta}_0 + \widehat{\beta}_1 \cdot 1 + \widehat{\beta}_2\,x_2 + \cdots + \widehat{\beta}_k\,x_k\right\}}{\exp\left\{\widehat{\beta}_0 + \widehat{\beta}_1 \cdot 0 + \cdots + \widehat{\beta}_k\,x_k\right\}} = \exp\left\{\widehat{\beta}_1\right\}.$$

Equivalently, the quantity $\exp\{\widehat{\beta}_1\} \cdot 100\%$ represents the estimated percent ratio of rates when $x_1 = 1$ and when $x_1 = 0$, while the other predictors are held constant.

5.1.4 Predicted Response

Taking into account(5.2), for a given set of predictors $x_1^0, \ldots, x_k^0$, the predicted response y^0 is computed as $y^0 = \exp\left\{\widehat{\beta}_0 + \widehat{\beta}_1\, x_1^0 + \cdots + \widehat{\beta}_k\, x_k^0\right\}$.

5.1.5 SAS Implementation

In SAS, the procedure `genmod` with the option `dist=poisson link=log` is used to fit a Poisson regression model.

5.1.6 R Implementation

In R, the function `glm()` with the option `family="poisson"`(link=log) fits a Poisson regression model.

5.1.7 Example

EXAMPLE 5.1. Number of days of hospital stay was recorded for 45 patients with chest pain, along with their gender, age, and history of chronic cardiac illness. We regress the number of days on the other variables via a Poisson regression model. To this end, we run the following SAS code:

```
data hospital_stay;
input days gender$ age illness$ @@;
cards;
1  F  31  yes   0  F  28  no    0  M  52  yes
1  M  72  yes   0  F  29  no    0  F  30  no
1  M  74  no    2  M  30  yes   2  F  72  no
1  M  58  no    2  F  28  no    2  F  65  no
2  M  65  no    1  M  52  no    4  M  51  no
2  F  63  no    0  F  31  no    1  F  47  yes
1  M  49  no    2  M  71  yes   2  M  48  no
2  F  47  no    0  F  31  no    3  M  44  yes
3  M  44  no    3  M  54  yes   4  F  72  yes
4  M  56  yes   3  F  73  yes   1  F  46  no
3  M  58  no    4  M  70  yes   2  M  36  no
1  M  50  no    1  M  59  no    0  M  52  no
6  M  68  yes   2  F  41  no    1  M  31  yes
1  M  69  no    3  M  73  no    3  F  77  yes
2  F  54  no    4  M  69  yes   5  M  68  yes
;
```

```
proc genmod;
class gender(ref='F') illness(ref='no');
   model days=gender age illness/dist=poisson link=log;
run;
```

The relevant output is:

Full Log Likelihood   -68.2139

Analysis Of Maximum Likelihood Parameter Estimates
Parameter		Estimate	Pr > ChiSq
Intercept		-0.8263	0.0789
gender	M	0.2264	0.3315
age		0.0205	0.0093
illness	yes	0.4477	0.0440

```
proc genmod data=hospital_stay;
   model days=/dist=poisson link=log;
run;
```

Full Log Likelihood   -77.0978

```
data deviance_test;
 deviance=-2*(-77.0978-(-68.2139));
   pvalue=1-probchi(deviance,3);
run;

proc print;
run;
```

deviance	pvalue
17.7678	0.000491110

In the fitted model, the estimated rate is $\widehat{\lambda} = \exp\{-0.8263 + 0.2264 * male + 0.0205 * age + 0.4477 * illness\}$. The p-value in the deviance test is less than 0.05, and so the model has a good fit. Patient's age and indicator of a chronic cardiac illness are significant predictors of average length of stay at the 5% significance level. For a one-year increase in patient's age, the estimated average number of days of hospital stay increases by $(\exp\{0.0205\} - 1) \cdot 100\% = 2.07\%$. Also, the estimated average number of days of hospital stay for patients with a chronic cardiac illness is $\exp\{0.4477\} \cdot 100\% = 156.47\%$ of that for patients without it.

The predicted length of stay for a 55-year old male with no chronic cardiac illness is computed as $y^0 = \exp\{-0.8263 + 0.2264 + 0.0205 * 55\} = 1.6949$. This value can be verified in SAS by running the following statements:

```
data prediction;
input gender$ age illness$;
cards;
M 55 no
;

data hospital_stay;
 set hospital_stay prediction;
run;

 proc genmod;
class gender(ref='F') illness(ref='no');
   model days=gender age illness
       /dist=poisson link=log;
 output out=outdata p=pred_days;
run;

proc print data=outdata(firstobs=46 obs=46);
var pred_days;
run;

pred_days
  1.69207
```

The R script and the relevant output are given below.

```
hospitalstay.data<-read.csv(file="./Example5.1Data.csv",
header=TRUE, sep=",")

#fitting Poisson model
summary(fitted.model<- glm(days ~ gender + age + illness,
data=hospitalstay.data, family=poisson(link=log)))
```

Coefficients:
                Estimate   Pr(>|z|)

```
(Intercept)   -0.826269   0.07888
genderM        0.226425   0.33145
age            0.020469   0.00931
illnessyes     0.447653   0.04404
```

```
#checking model fit
intercept.only.model<- glm(days ~ 1,
data=hospitalstay.data, family=poisson(link=log))
print(deviance<- -2*(logLik(intercept.only.model)
-logLik(fitted.model)))
```

17.76773

```
print(p.value<- pchisq(deviance, df=3, lower.tail=FALSE))
```

0.0004911253

```
#using fitted model for prediction
print(predict(fitted.model, data.frame(gender="M",
age=55, illness="no"), type="response"))
```

1.692066

$\square$

5.2   Zero-truncated Poisson Regression Model

5.2.1   Model Definition

If the response variable y assumes only positive integer values (no zeros), then the data may be modeled by means of a *zero-truncated Poisson regression model*[2]. Let $x_1, \ldots, x_k$ be the predictors in this model. Then the response variable y follows a zero-truncated Poisson distribution with the probability mass function

$$\mathbb{P}(Y = y) = \frac{\lambda^y \exp\{-\lambda\}}{y! \left(1 - \exp\{-\lambda\}\right)}, \quad y = 1, 2, \ldots,$$

where

$$\lambda = \exp\left\{\beta_0 + \beta_1 x_1 + \cdots + \beta_k x_k\right\}. \tag{5.3}$$

---

[2]Introduced in Gurmu, S. (1991). "Tests for detecting overdispersion in the positive Poisson regression model". *Journal of Business and Economic Statistics*, 9(2): 215 – 222.

For a zero-truncated Poisson distribution, the expected value of y is (show it!)

$$\mathbb{E}(y) = \frac{\lambda}{1 - \exp\{-\lambda\}}, \tag{5.4}$$

and, even though, the probability mass function belongs to the exponential family of distributions, this regression is not a generalized linear model because the log link function in (5.3) relates linear regression to λ which is not the expected value of y.

5.2.2 Fitted Model

In a fitted zero-truncated Poisson regression model, according to (5.3), the estimate of λ is written as

$$\widehat{\lambda} = \exp\left\{\widehat{\beta}_0 + \widehat{\beta}_1 x_1 + \cdots + \widehat{\beta}_k x_k\right\}. \tag{5.5}$$

5.2.3 Interpretation of Estimated Regression Coefficients

The formula for the expected value of y (5.4) contains an exponentiated λ. Since by (5.3) lambda itself is an exponential function, this expected value may be assumed negligibly different from λ for most values of λ. Hence, the way the estimated regression coefficients are interpreted in the fitted Poisson regression model (see Subsection 5.1.3) remains in effect.

5.2.4 Predicted Response

In view of (5.4) and (5.5), for some given values $x_1^0, \ldots, x_k^0$, the predicted response y^0 is found as

$$y^0 = \frac{\exp\left\{\widehat{\beta}_0 + \widehat{\beta}_1 x_1^0 + \cdots + \widehat{\beta}_k x_k^0\right\}}{1 - \exp\left\{-\exp\left\{\widehat{\beta}_0 + \widehat{\beta}_1 x_1^0 + \cdots + \widehat{\beta}_k x_k^0\right\}\right\}}.$$

5.2.5 SAS Implementation

The procedure `fmm` (stands for "finite mixture models") with the option `dist=truncpoisson` in the model statement may be used to fit a zero-truncated Poisson model. The following syntax invokes the procedure:

```
proc fmm data=data_name;
    class <list of categorical predictors>;
        model response_name=<list of predictors>/dist=truncpoisson;
run;
```

For categorical predictors, the level that comes last in alphabetical order is used for reference. The `fmm` procedure doesn't allow specification of reference categories in the `class` statement. It can be done by changing the format of categorical variables in a `proc format` statement.

5.2.6 R Implementation

In R, the function `vglm()` (stands for "vector generalized linear models") in the library `VGAM` is used to fit a zero-truncated Poisson model. The general form of this function is

summary($fitted.model.name$<- vglm($response.name \sim x1.name + \cdots$ + $xk.name$, data=$data.name$, family=pospoisson()))

• The function `pospoisson()` fits a positive Poisson distribution, that is, the zero-truncated Poisson.

5.2.7 Example

EXAMPLE 5.2. Consider the setting in Example 5.1. Suppose investigators are not concerned with outpatients, that is, those who were treated and dismissed the same day. Hence, the data are reduced to the 38 patients who spent at least one day in the hospital. We run a zero-truncated Poisson model with the three predictor variables along with the intercept-only model to check the fit. The SAS code and respective outputs are:

```
data hospital_days;
  set hospital_stay;
if (days>0);
run;

proc format;
value $genderfmt 'F'='ref' 'M'='M';
value $illnessfmt 'yes'='illness' 'no'='ref';
run;
```

```
proc fmm;
class gender illness;
 model days=gender age illness/dist=truncpoisson;
format gender $genderfmt. illness $illnessfmt.;
run;
```

-2 Log Likelihood 105.2

Parameter Estimates for Truncated Poisson Model

Effect	gender	illness	Estimate	Pr > |z|
Intercept			-0.7041	0.2797
gender	M		0.2146	0.4559
gender	ref		0	.
age			0.01604	0.1056
illness		illness	0.5903	0.0296
illness		ref	0	.

```
proc fmm;
 model days=/dist=truncpoisson;
run;
```

-2 Log Likelihood 115.2

```
data deviance_test;
 deviance=115.2-105.2;
   pvalue=1-probchi(deviance,3);
run;
```

```
proc print;
run;
```

deviance	pvalue
10	0.018566

In the fitted model, the estimated rate is $\widehat{\lambda} = \exp\{-0.7041 + 0.2146 * male + 0.01604 * age + 0.5903 * illness\}$. The p-value in the goodness-of-fit test is smaller than 0.05, confirming that the model fits the data well. Only presence of chronic cardiac illness is significant at the 5% level. The estimated average number of days of hospital stay for patients with a chronic cardiac illness is $\exp\{0.5903\} \cdot 100\% = 180.45\%$ of that for patients without a it.

To predict the number of days of hospital stay for a 55-year old male without a chronic cardiac illness, we calculate

$$y^0 = \frac{\exp\left\{-0.7041 + 0.2146 + 0.01604 * 55\right\}}{1 - \exp\left\{-\exp\left\{-0.7041 + 0.2146 + 0.01604 * 55\right\}\right\}} = 1.9169.$$

A similar result may be obtained from SAS by running these lines of code:

```
data prediction;
input gender$ age illness$;
cards;
M 55 no
;

data hospital_days;
 set hospital_days prediction;
run;

proc fmm ;
class gender illness;
 model days=gender age illness/dist=truncpoisson;
 output out=outdata pred=pred_days;
format gender $genderfmt. illness $illnessfmt.;
run;

proc print data=outdata(firstobs=39 obs=39);
var pred_days;
run;

pred_days
  1.91706
```

R script and output for this example are presented below.

```
hospitalstay.data<- read.csv(file="./Example5.1Data.csv",
header=TRUE, sep=",")

#eliminating zeros from the original data set
hospitaldays.data<- hospitalstay.data
[which(hospitalstay.data$days!=0),]
```

```
install.packages("VGAM")
library(VGAM)

#fitting zero-truncated Poisson model
summary(fitted.model<- vglm(days ~ gender + age + illness,
data=hospitaldays.data, family=pospoisson()))
```

Coefficients:

	Estimate	Pr(>|z|)
(Intercept)	-0.704061	0.2797
genderM	0.214611	0.4558
age	0.016042	0.1056
illnessyes	0.590345	0.0296

```
#checking model fit
intercept.only.model<- vglm(days ~ 1,
data=hospitaldays.data, family=pospoisson())
print(deviance<- -2*(logLik(intercept.only.model)
-logLik(fitted.model)))
```

9.993755

```
print(p.value<- pchisq(deviance, df=3, lower.tail=FALSE))
```

0.0186193

```
#using fitted model for prediction
print(predict(fitted.model, data.frame(gender="M",
age=55, illness="no"), type="response"))
```

1.917057

□

5.3   Zero-inflated Poisson Regression Model

5.3.1   Model Definition

Suppose that one of the variables recorded during a health survey was the number of cigarettes the respondent smoked yesterday. Some respondents may have reported zero number of cigarettes smoked. There are two possible scenarios: either the respondents do not smoke at all, or they happened not to smoke a single cigarette that day. That is, the observed zero may be either a *structural zero*, when the respondent's behavior lies outside of the behavioral repertoire under study (for example, a person doesn't smoke), or a *chance zero*, when the respondent's typical behavior falls within the behavioral range under study, but just not during a particular time period (in this case, a person normally smokes at least one cigarette a day, but just happened not to smoke yesterday).

The presence of structural zeros inflates the number of zeros in the Poisson model, which makes the Poisson model invalid, and a *zero-inflated Poisson model* (often abbreviated as ZIP) should be used instead. The ZIP model attempts to separate the structural zeros from the chance zeros, by viewing the response variable y as assuming the value zero with probability π (the case of a structural zero), and otherwise, with probability $1 - \pi$, being a count variable with the Poisson distribution with rate λ.

To increase the rigor, in the *zero-inflated Poisson regression model*[3] with the predictor variables $x_1, \ldots, x_k$, the response variable y has a probability distribution defined as follows:

$$\mathbb{P}(Y = y) = \begin{cases} \pi + (1 - \pi) \exp\{-\lambda\}, & \text{if } y = 0, \\ (1 - \pi) \dfrac{\lambda^y \exp\{-\lambda\}}{y!}, & \text{if } y = 1, 2, \ldots, \end{cases} \tag{5.6}$$

where

$$\pi = \frac{\exp\{\beta_0 + \beta_1 x_1 + \cdots + \beta_m x_m\}}{1 + \exp\{\beta_0 + \beta_1 x_1 + \cdots + \beta_m x_m\}}, \tag{5.7}$$

and

$$\lambda = \exp\left\{\gamma_0 + \gamma_1 x_{m+1} + \cdots + \gamma_{k-m} x_k\right\}. \tag{5.8}$$

Here the first m of the predictor variables are being used to model the probability of a structural zero π, while the rest of the x variables are used as predictors for the Poisson rate λ. The β and γ coefficients are the parameters of this model.

The zero-inflated Poisson distribution defined in (5.6) is a mixture of two distributions: a Poisson and a point mass at zero. This type of distribution is not a representative of the exponential family of distributions. Consequently, the zero-inflated Poisson model is not a member of the class of generalized linear models.

---

[3]Introduced in Lambert, D. (1992). "Zero-inflated Poisson regression, with an application to defects in manufacturing". *Technometrics*, 34(1): 1 – 14.

5.3.2 Fitted Model

By (5.7) and (5.8), in the fitted zero-inflated Poisson regression model, the estimated parameters are

$$\widehat{\pi} = \frac{\exp\{\widehat{\beta}_0 + \widehat{\beta}_1 x_1 + \cdots + \widehat{\beta}_m x_m\}}{1 + \exp\{\widehat{\beta}_0 + \widehat{\beta}_1 x_1 + \cdots + \widehat{\beta}_m x_m\}}, \tag{5.9}$$

and

$$\widehat{\lambda} = \exp\left\{\widehat{\gamma}_0 + \widehat{\gamma}_1 x_{m+1} + \cdots + \widehat{\gamma}_{k-m} x_k\right\}. \tag{5.10}$$

5.3.3 Interpretation of Estimated Regression Coefficients

Since in the zero-inflated Poisson model the probability π is modeled via a logistic link function, the interpretation of the estimated beta coefficients is identical to those in a binary logistic regression model (see Subsection 3.1.3). Also, the expected value of the response variable y is $\mathbb{E}(y) = (1 - \pi)\lambda$ (see Exercise 5.7), and the sets of x variables in the definitions of π and λ are chosen to be non-overlapping. This implies that when we are interpreting the estimated gamma coefficients, we may assume that π is fixed, and thus the interpretation is the same as in the Poisson regression model (see Subsection 5.1.3). Note that it is possible to use the same x variables in the regression parts of π and λ, but the estimates of the regression coefficients won't be easily interpretable.

5.3.4 Predicted Response

In view of the above expression for $\mathbb{E}(y)$ and formulas (5.9) and (5.10), when $x_1^0, x_2^0, \ldots, x_k^0$ are fixed, the predicted response y^0 is computed as

$$y^0 = \left(1 - \frac{\exp\{\widehat{\beta}_0 + \widehat{\beta}_1 x_1^0 + \cdots + \widehat{\beta}_m x_m^0\}}{1 + \exp\{\widehat{\beta}_0 + \widehat{\beta}_1 x_1^0 + \cdots + \widehat{\beta}_m x_m^0\}}\right) \exp\left\{\widehat{\gamma}_0 + \widehat{\gamma}_1 x_{m+1}^0 + \cdots + \widehat{\gamma}_{k-m} x_k^0\right\}. \tag{5.11}$$

5.3.5 SAS Implementation

The ZIP model may be requested in SAS by adding an option `dist=zip` to the `model` statement of the `genmod` procedure, and including the `zeromodel` statement. The general syntax is

```
proc genmod data=data_name;
   class <list of categorical predictors>;
      model response_name = <list of predictors>/dist=zip;
         zeromodel <list of predictors of structural zeros>;
run;
```

5.3.6   R Implementation

Function `zeroinfl()` in the library `pscl`, which stands for "Political Science Computational Laboratory", is used to fit a zero-inflated Poisson model in R. The general form of this function is

summary(*fitted.model.name*<- `zeroinfl`(*response.name* ~ $x\{m{+}1\}$.*name* + $\cdots$ + *xk.name*|*x1.name* + $\cdots$ + *xm.name*, `data` = *data.name*))

5.3.7   Example

EXAMPLE 5.3. A health survey was been administered to a random sample of 40 people aged between 25 and 50. Their gender, self-reported health condition (excellent or good), age, and the number of cigarettes they smoked yesterday were recorded. Since those respondents who don't smoke were included in the survey, it is expected that the number of cigarettes smoked would have a Poisson distribution with inflated number of zeros. Below we fit a ZIP model where health condition is used as the predictor of structural zero, while gender and age are the count model predictors. The SAS code and output are:

```
data smoking;
   input gender$ health$ age cigarettes @@;
   cards;
M good    34 3  F exclnt 48 1  M exclnt 26 0  M good    39 0
F good    27 1  M good    28 5  F good    44 1  M exclnt 30 0
F exclnt 26 0  F good    38 2  F good    40 1  F exclnt 31 0
M good    27 3  F exclnt 34 1  F good    36 2  F exclnt 34 2
F exclnt 39 0  F good    42 1  F good    48 4  M good    32 5
M good    47 2  M good    29 3  M exclnt 38 0  F good    50 4
M good    30 3  M good    38 2  M good    31 6  F exclnt 33 0
F good    28 0  F good    42 3  M exclnt 28 0  M good    31 2
F exclnt 31 0  F exclnt 42 0  F good    44 4  F good    39 1
```

118

M good   40 6 M good   39 3 M exclnt 25 0 F good   45 2
;

```
proc genmod;
class gender(ref='F') health(ref='good');
 model cigarettes=gender age/dist=zip;
  zeromodel health;
run;
```

Full Log Likelihood -57.0406

Analysis Of Maximum Likelihood Parameter Estimates

Parameter	Estimate	Pr > ChiSq
Intercept	-0.1381	0.8692
gender M	0.7268	0.0107
gender F	0.0000	.
age	0.0186	0.3509

Analysis Of Maximum Likelihood Zero Inflation Parameter Estimates

Parameter	Estimate	Pr > ChiSq
Intercept	-3.7950	0.0876
health exclnt	4.9195	0.0341
health good	0.0000	.

```
proc genmod;
  model cigarettes=/dist=zip;
    zeromodel;
run;
```

Full Log Likelihood -71.3892

```
data deviance_test;
 deviance=-2*(-71.3892-(-57.0406));
   pvalue=1-probchi(deviance,3);
run;

proc print;
run;
```

deviance	pvalue
28.6972	0.000002593

From this output, the fitted regression model has estimated parameters

$$\widehat{\pi} = \frac{\exp\{-3.7950 + 4.9195 * excellent\_health\}}{1 + \exp\{-3.7950 + 4.9195 * excellent\_health\}},$$

and

$$\widehat{\lambda} = \exp\{-0.1381 + 0.0186 * age + 0.7268 * male\}.$$

This model has a good fit, since the p-value in the deviance test is very small. At the 5% level, health status significantly predicts the odds of being a non-smoker, whereas gender has a significant effect on the average number of cigarettes smoked in a day. As follows from these estimates, the estimated odds of not smoking for people in excellent health is $\exp\{4.9195\} \cdot 100\% = 13,694.26\%$ of those for people in good health. Also, the estimated average number of cigarettes smoked in a day by men is $\exp\{0.7268\} \cdot 100\% = 206.85\%$ of the number of cigarettes smoked by women.

Further, by (5.11), the predicted number of cigarettes smoked per day by a 50-year old male who is in good health is found as

$$y^0 = \left(1 - \frac{\exp\{-3.7950\}}{1 + \exp\{-3.7950\}}\right) \exp\{-0.1381 + 0.0186 * 50 + 0.7268\} = 4.4659.$$

The SAS code for prediction is

```
data prediction;
input gender$ health$ age;
cards;
M good 50
;

data smoking;
 set smoking prediction;
run;

proc genmod;
class gender(ref='F') health(ref='good');
 model cigarettes=gender age/dist=zip;
   zeromodel health;
    output out=outdata p=pred_cig;
run;

proc print data=outdata(firstobs=41 obs=41);
var pred_cig;
run;
```

pred_cig
   4.47333

R script and output for this example follow.

```
smoking.data<-read.csv(file="./Example5.3Data.csv",
header=TRUE, sep=",")
install.packages("pscl")
library(pscl)

#specifying reference category
health.rel<- relevel(smoking.data$health, ref="good")

#fitting zero-inflated Poisson model
summary(fitted.model<- zeroinfl(cigarettes ~ gender
 + age|health.rel, data=smoking.data))
```

Count model coefficients (poisson with log link):

	Estimate	Pr(>|z|)
(Intercept)	-0.13820	0.8690
genderM	0.72686	0.0107
age	0.01863	0.3506

Zero-inflation model coefficients (binomial with logit link):

	Estimate	Pr(>|z|)
(Intercept)	-3.795	0.0875
health.relexclnt	4.920	0.0341

```
#checking model fit
intercept.only.model<- zeroinfl
(cigarettes ~ 1, data=smoking.data)
print(deviance<- -2*(logLik(intercept.only.model)
-logLik(fitted.model)))
```

28.6972

```
print(p.value<- pchisq(deviance, df=3, lower.tail=FALSE))
```

2.592711e-06

```
#using fitted model for prediction
print(predict(fitted.model, data.frame(gender="M",
health.rel="good", age=50)))
```

4.473475

□

5.4 Hurdle Poisson Regression Model

5.4.1 Model Definition

The hurdle Poisson regression is applied when the response variable y has a Poisson distribution with an inflated number of zeros, and, moreover, there is a reason to believe that the underlying characteristics of cases with zeros systematically differ from those with positive responses. Put differently, zeros are modeled separately from zero-truncated Poisson observations. Thus, for predictors $x_1, \ldots, x_k$, the *hurdle Poisson regression model*[4] assumes that the response variable y has a probability distribution function

$$\mathbb{P}(Y = y) = \begin{cases} \pi, & \text{if } y = 0, \\ (1 - \pi) \dfrac{\lambda^y \exp\{-\lambda\}}{y!(1 - \exp\{-\lambda\})}, & \text{if } y = 1, 2, \ldots, \end{cases} \tag{5.12}$$

where

$$\pi = \frac{\exp\{\beta_0 + \beta_1 x_1 + \cdots + \beta_m x_m\}}{1 + \exp\{\beta_0 + \beta_1 x_1 + \cdots + \beta_m x_m\}}, \tag{5.13}$$

and

$$\lambda = \exp\left\{\gamma_0 + \gamma_1 x_{m+1} + \cdots + \gamma_{k-m} x_k\right\}. \tag{5.14}$$

The distribution of y is a mixture of a degenerate distribution at zero and a zero-truncated Poisson distribution. This type of distribution is not in the exponential family, and thus, the hurdle Poisson model is not a generalized linear model.

5.4.2 Fitted Model

By (5.13) and (5.14), in a fitted hurdle Poisson regression model, estimated parameters are

$$\widehat{\pi} = \frac{\exp\{\widehat{\beta}_0 + \widehat{\beta}_1 x_1 + \cdots + \widehat{\beta}_m x_m\}}{1 + \exp\{\widehat{\beta}_0 + \widehat{\beta}_1 x_1 + \cdots + \widehat{\beta}_m x_m\}} \tag{5.15}$$

and

$$\widehat{\lambda} = \exp\left\{\widehat{\gamma}_0 + \widehat{\gamma}_1 x_{m+1} + \cdots + \widehat{\gamma}_{k-m} x_k\right\}. \tag{5.16}$$

---

[4]Originally examined in Mullahy, J. (1986). "Specification and testing of some modified count data models". *Journal of Econometrics*, 33(3): 341 – 365.

5.4.3 Interpretation of Estimated Regression Coefficients

It can be shown (see Exercise 5.11) that the expected value of the response variable $\mathbb{E}(y) = (1-\pi)\dfrac{\lambda}{1 - \exp\{-\lambda\}}$, and thus, estimated regression coefficients in π and λ are interpreted as in a binary logistic (see Subsection 3.1.3) and Poisson regressions (see Subsection 5.1.3), respectively, provided all the other predictors remain the same.

5.4.4 Predicted Response

Taking into account the above expression for the expected value of y and relations (5.15) and (5.16), we can write the predicted response as

$$y^0 = \left(1 - \frac{\exp\{\widehat{\beta}_0 + \widehat{\beta}_1\, x_1^0 + \cdots + \widehat{\beta}_m\, x_m^0\}}{1 + \exp\{\widehat{\beta}_0 + \widehat{\beta}_1\, x_1^0 + \cdots + \widehat{\beta}_m\, x_m^0\}}\right) \times$$

$$\times \frac{\exp\left\{\widehat{\gamma}_0 + \widehat{\gamma}_1\, x_{m+1}^0 + \cdots + \widehat{\gamma}_{k-m}\, x_k^0\right\}}{1 - \exp\{-\exp\left\{\widehat{\gamma}_0 + \widehat{\gamma}_1\, x_{m+1}^0 + \cdots + \widehat{\gamma}_{k-m}\, x_k^0\right\}\}}.$$

5.4.5 SAS Implementation

The hurdle Poisson model may be fit by means of the **fmm** procedure with the syntax below.

```
proc fmm data=data_name;
   class <list of categorical predictors>;
     model response_name = <list of predictors>/dist=truncpoisson;
       model + /dist=constant;
   probmodel <list of predictors of zeros>;
run;
```

• The second **model** statement adds a degenerate distribution at point zero.
• The **probmodel** statement fits the probability of zero that, in our notation, is equal to $1 - \pi$. Applying (5.13), we obtain

$$1 - \pi = 1 - \frac{\exp\{\beta_0 + \beta_1\, x_1 + \cdots + \beta_m\, x_m\}}{1 + \exp\{\beta_0 + \beta_1\, x_1 + \cdots + \beta_m\, x_m\}}$$

$$= \frac{\exp\{-(\beta_0 + \beta_1\, x_1 + \cdots + \beta_m\, x_m)\}}{1 + \exp\{-(\beta_0 + \beta_1\, x_1 + \cdots + \beta_m\, x_m)\}}.$$

It means that when fitting the model using our notation, the signs of estimated beta regression coefficients have to be reversed.

5.4.6 R Implementation

In R, the function `hurdle()` in the library `pscl` may be used to fit a hurdle Poisson regression model. The general syntax is

summary(*fitted.model.name*<- hurdle(*response.name* $\sim$ $x\{m+1\}.name$ + $\cdots$ + *xk.name* | *x1.name* + $\cdots$ + *xm.name*, data = *data.name*, dist="poisson", zero.dist="binomial", link="logit"))

5.4.7 Example

EXAMPLE 5.4. A college bookstore wants to gain insight into the textbook purchasing habits of students. A random sample of 40 students was drawn, and it was recorded how many textbooks each student purchased through the bookstore for the current term, whether the student is currently renting any textbooks, and whether a student has a financial aid. The manager at the bookstore conjectures that no textbook purchase might be attributed to students' renting textbooks, and that students with financial aid tend to purchase more textbooks. The data and the analysis follow.

```
data bookstore;
input ntextbooks renting aid$ @@;
cards;
0 3 no   0 2 no   3 0 yes   0 0 no   0 1 no   1 0 no
0 3 no   2 0 yes 4 0 no   0 2 no   4 1 no   7 0 yes
3 2 yes 0 4 no   1 2 no   0 0 no   0 5 no   1 0 no
0 2 no   3 0 no   0 3 no   6 1 yes 2 1 yes 1 0 no
6 0 yes 2 0 no   0 3 no   4 0 yes 0 1 no   0 2 no
3 0 no   3 2 no   3 0 yes   2 0 yes 0 3 no   2 1 no
3 0 yes 1 3 no   3 0 yes   0 2 no

proc format;
value $aidfmt 'no'='ref' 'yes'='aid';
run;

proc fmm;
class aid;
 model ntextbooks=aid/dist=truncpoisson;
   model+/dist=constant;
  probmodel renting;
format aid $aidfmt.;
```

```
run;
```

-2 Log Likelihood 115.7

Parameter Estimates for Truncated Poisson Model

Effect	aid	Estimate	Pr > |z|
Intercept		0.5951	0.0147
aid	aid	0.6754	0.0204
aid	ref	0	.

Parameter Estimates for Mixing Probabilities

Effect	Estimate	Pr > |z|
Intercept	2.0494	0.0016
renting	-1.2749	0.0009

```
proc fmm;
 model ntextbooks=/dist=truncpoisson;
  model + /dist=constant;
   probmodel;
run;
```

-2 Log Likelihood 139.1

```
data deviance_test;
 deviance=139.1-115.7;
   pvalue=1-probchi(deviance,2);
run;

proc print;
run;
```

deviance	pvalue
23.4	0.000008294

Reversing the estimated beta coefficients, we obtain that in the fitted model, the parameter estimates are $\widehat{\pi} = \dfrac{\exp\{-2.0494 + 1.2749 * renting\}}{1 + \exp\{-2.0494 + 1.2749 * renting\}}$, and $\widehat{\lambda} = \exp\{0.5951 + 0.6754 * aid\}$. The model fits the data well since the p-value in the deviance test is tiny. Whether a student rents textbooks significantly predicts the probability of not purchasing them, whereas presence of financial aid is significantly associated with the mean number of purchased textbooks. If the number of rented books increases by one, the estimated odds in favor of not buying textbooks increases by $(\exp\{1.2749\} - 1) \cdot 100\% = 257.83\%$. The estimated average number of purchased textbooks for a student with a financial aid is exp{0.6754} · 100% = 196.48% of that for a student without it.

Prediction of the number of textbooks purchased by a student who has no rented books or financial aid is calculated as follows:

$$y^0 = \left(1 - \frac{\exp\{-2.0494\}}{1 + \exp\{-2.0494\}}\right) \cdot \frac{\exp\left\{0.5951\right\}}{1 - \exp\{-\exp\{0.5951\}\}} = 1.9194.$$

SAS produces similar prediction as can be seen by running the statements below.

```
data prediction;
input renting aid$;
cards;
0 no
;

data bookstore;
 set bookstore prediction;
run;

proc fmm;
class aid;
  model ntextbooks=aid/dist=truncpoisson;
   model+/dist=constant;
 probmodel renting;
  output out=outdata pred=p_ntextbooks;
format aid $aidfmt.;
run;

proc print data=outdata(firstobs=41 obs=41);
var p_ntextbooks;
run;

p_ntextbooks
    1.91942
```

The R script and output are:

```
bookstore.data<-read.csv(file="./Example5.4Data.csv",
header=TRUE, sep=",")
install.packages("pscl")
```

```
library(pscl)

#fitting hurdle Poisson model
summary(fitted.model<- hurdle(ntextbooks ~ aid|renting,
data=bookstore.data, dist="poisson", zero.dist="binomial",
link="logit"))
```

Count model coefficients (truncated poisson with log link):

	Estimate	Pr($>$|z|)
(Intercept)	0.5951	0.0147
aidyes	0.6754	0.0204

Zero hurdle model coefficients (binomial with logit link):

	Estimate	Pr($>$|z|)
(Intercept)	2.0494	0.001610
renting	-1.2749	0.000923

```
#checking model fit
intercept.only.model<- hurdle(ntextbooks ~ 1,
data=bookstore.data,
dist="poisson", zero.dist="binomial", link="logit")
print(deviance<- -2*(logLik(intercept.only.model)
-logLik(fitted.model)))
```

23.39548

```
print(p.value<- pchisq(deviance, df=2, lower.tail=FALSE))
```

8.312594e-06

```
#using fitted model for prediction
print(predict(fitted.model, data.frame(renting=0, aid="no")))
```

1.919422

□

Exercises for Chapter 5

EXERCISE 5.1. The number of defective items produced by a machine operator during one shift is modeled through a Poisson regression where independent variables are the length of work experience as a machine operator (in years), and whether it was a morning, day, evening, or night shift. The data were obtained for 36 randomly chosen shifts and operators.

Defctv	Expr	Shift	Defctv	Expr	Shift
2	3.1	morning	0	2.1	day
5	2.1	morning	2	3.0	day
3	8.0	morning	5	8.2	evening
3	7.6	morning	4	4.0	evening
2	5.9	morning	4	6.2	evening
2	4.0	morning	3	2.9	evening
1	1.7	morning	2	2.1	evening
0	1.8	morning	2	1.9	evening
0	8.2	morning	1	6.7	evening
1	8.1	morning	1	3.4	evening
3	3.0	day	1	7.6	evening
3	7.7	day	6	5.1	night
2	6.3	day	4	3.2	night
2	8.1	day	4	7.6	night
2	7.7	day	4	2.5	night
1	2.4	day	3	6.2	night
1	3.0	day	3	2.0	night
1	4.6	day	5	4.0	night

(a) Run the Poisson regression model. Discuss the significance of predictors at the 5% level of significance.
(b) Write down the estimated model. How good is the fit of the model?
(c) Give interpretation of the estimated significant coefficients.
(d) Predict the number of defective items produced during a night shift by an operator with six months of experience.

EXERCISE 5.2. A large automobile insurance company is studying the relation between the total number of auto accidents (including minor) that a policyholder had caused, and the policyholder's gender, age, and total number of miles driven (in thousands). The data for 45 randomly chosen policyholders are given in the table below.

Accid	Gender	Age	Miles	Accid	Gender	Age	Miles
1	M	27	90	4	F	32	180
1	M	60	70	3	F	29	180
5	M	20	25	4	F	51	90
3	M	42	75	0	F	40	190
3	M	55	170	4	F	43	90
5	M	67	160	4	F	43	20
2	M	32	80	1	F	47	160
3	M	42	70	0	F	36	190
4	M	36	120	4	F	31	120
3	M	30	110	5	F	40	100
2	M	27	150	4	F	50	130
5	M	33	140	3	F	51	150
9	M	68	180	6	F	48	170
5	M	41	50	1	F	59	70
5	M	43	150	0	F	57	140
6	M	59	130	6	F	57	180
7	M	65	90	2	F	40	170
2	M	58	150	4	F	36	50
3	M	54	170	3	F	49	150
0	F	33	110	2	F	58	60
2	F	44	170	1	F	55	180
2	F	36	100	8	F	66	130
3	F	53	200				

(a) Fit the Poisson model to the data and specify estimated parameters. What variables are statistically significant predictors of the number of car accidents? Use $\alpha = 0.05$.

(b) Check goodness-of-fit of the model.

(c) Interpret the estimated significant regression coefficients.

(d) Give a predicted value of the total number of auto accidents caused by a 35-year-old woman who has driven a total of one hundred thousand miles.

EXERCISE 5.3. A howling survey is a productive method for estimating the minimum number of wolves within a pack. Thirty two surveys have been conducted at one site per 100 acres of land. The count of individual wolves that called back, time of howling session (dusk or night) and wind speed (in mph) were recorded. Also, presence of source of drinking water was noted. The data are summarized in the table below.

Calls	Time	Wind	Water	Calls	Time	Wind	Water
2	dusk	0	yes	3	dusk	0	yes
2	dusk	1	yes	0	dusk	3	no
3	dusk	0	no	1	dusk	3	yes
2	night	6	no	2	dusk	3	yes
3	dusk	2	no	7	night	2	yes
4	night	3	yes	5	dusk	0	yes
5	dusk	1	yes	2	night	0	yes
3	night	5	yes	4	night	2	no
4	night	5	yes	6	night	1	yes
7	night	0	yes	3	night	3	yes
1	dusk	6	yes	0	dusk	1	no
2	night	1	no	1	dusk	3	no
4	dusk	2	yes	4	night	3	yes
6	night	2	yes	1	dusk	0	yes
5	dusk	3	yes	4	dusk	2	yes
2	night	3	yes	1	dusk	2	yes

(a) Fit a Poisson regression model for the number of calls. Discuss the model fit.

(b) Specify the fitted model. Give estimates of all parameters. Which variables are significant at the 5%?

(c) Give interpretation of estimated significant regression coefficients.

(d) What is the predicted number of wolves that would call back during a howling session conducted at dusk, in a wilderness with no water source, if the wind's speed is 5 mph?

EXERCISE 5.4. Reduce the data in Exercise 5.1 to only those operators who produced defective items.

(a) Model the number of defective items via the zero-truncated Poisson regression model. Display the fitted model. List the significant predictors.

(b) Discuss the model fit.

(c) Interpret estimated significant coefficients.

(d) Predict the number of defective items produced during a night shift by an operator with six months of experience.

EXERCISE 5.5. In the setting of Exercise 5.2, remove those policyholders who caused no accidents. Run the zero-truncated Poisson regression model on the remaining data.

(a) Write down the fitted model. Are there any significant predictors at the 5% level?

(b) Discuss the fit of the model.

(c) Interpret the estimated significant beta coefficients.

(d) Give a predicted value of the total number of auto accidents caused by a 35-year-old woman who has driven a total of one hundred thousand miles.

EXERCISE 5.6. Trim the data in Exercise 5.3, leaving the records of the howling sessions when wolves were present in the area and responded.

(a) Model the number of wolves through a zero-truncated Poisson regression model. Estimate all parameters. Are there any significant predictors at the 0.05 level?

(b) Test the goodness-of-fit of the model.

(c) Give interpretation of the estimated significant regression coefficients.

(d) Find the predicted number of wolves that would call back during a howling session conducted at dusk, in a wilderness with no water source, if the wind's speed is 5 mph.

EXERCISE 5.7. Consider the zero-inflated Poisson regression model defined by (5.6) – (5.8).

(a) Show that the expected value of y is $\mathbb{E}(y) = (1 - \pi)\lambda$.

(b) Prove that the estimated gamma coefficients in the expression for $\widehat{\lambda}$ yield the same interpretation as in the Poisson regression model (see Subsection 5.1.3). Hint: use the fact that the predictors in the definitions of λ and π are different.

EXERCISE 5.8. On the day of a race, runners were asked how many races they participated in during the past two months. The runners' bib numbers were also noted, and after the race the data were obtained on the runners' gender, age, type of run on that day (5K/10K/full marathon). In addition, the average pace was calculated for each runner (in minutes per mile) from the distance and time of the runs. The data on 36 runners are

NRuns	Gender	Age	Run	Pace	NRuns	Gender	Age	Run	Pace
0	F	33	10K	10.04	1	F	51	5K	12.28
4	M	26	Full	7.17	1	F	35	10K	6.98
0	M	32	10K	11.14	2	M	25	10K	12.01
2	F	27	5K	9.18	3	M	34	5K	6.78
0	M	38	5K	7.52	0	M	28	5K	11.66
1	F	47	10K	11.59	0	F	39	10K	12.31
1	M	51	5K	9.44	2	M	32	Full	6.58
1	F	49	5K	9.53	3	F	44	Full	7.46
0	M	54	10K	8.48	0	F	49	10K	11.11
2	F	27	5K	11.71	2	M	52	Full	9.2
2	M	24	10K	7.56	1	M	30	5K	6.41
0	F	54	5K	13.78	1	M	43	10K	7.7
3	M	35	Full	7.34	1	M	30	10K	10.01
0	M	50	5K	7.51	0	M	53	5K	7.56
0	M	44	5K	8.92	2	F	46	Full	8.34
1	F	37	5K	10.71	0	F	28	5K	9.67
0	M	54	5K	8.72	2	F	50	Full	10.07
2	F	51	10K	7.41	2	F	54	5K	7.58

(a) Fit the zero-inflated Poisson regression to model the number of runs in the previous two months. Check if the type of today's run is significantly associated with inflation of zeros. Write down the fitted model.

(b) Discuss the model fit.

(c) Interpret the estimated significant coefficients.

(d) Calculate the predicted number of previous runs for a female runner, aged 45, who ran at an average pace of 10 minutes per mile, if she ran 10K.

EXERCISE 5.9. In an elementary school, children participate in a home reading club. They are asked to read at least 15 minutes every day and submit at the end of each month a list of books read. Some teachers make this assignment a part of the homework, others leave it as optional. A random sample of 50 students is drawn and the following variables recorded: grade level, whether reading assignment was a part of homework (yes/no), gender, and the number of books read at the proper grade level or above. The books below the grade level were not counted. If a student didn't turn in the list of books, then a zero was recorded for the number of books. The data are given below.

Grade	HW	Gender	Books	Grade	HW	Gender	Books
3	no	M	3	2	yes	F	0
3	yes	M	3	2	yes	M	3
2	no	F	4	2	no	M	3
2	yes	M	3	3	yes	F	4
3	no	F	2	3	yes	M	3
1	yes	F	0	3	yes	F	1
1	yes	F	4	1	no	M	0
2	no	F	0	2	no	M	0
1	no	M	0	1	yes	M	0
3	no	M	1	2	yes	F	6
3	yes	F	3	2	yes	F	2
2	no	F	4	2	no	F	3
3	no	M	0	2	no	F	0
2	no	M	0	3	no	F	5
1	yes	F	5	3	yes	M	2
3	yes	M	2	1	no	M	0
1	no	F	1	3	no	F	2
3	no	F	4	2	yes	F	0
1	no	F	0	2	no	M	2
2	yes	F	2	2	no	M	0
3	no	F	4	3	no	F	3
1	no	M	2	1	yes	F	1
2	no	M	0	2	no	F	0
2	no	F	4	1	yes	M	1
3	no	F	5	2	yes	M	2

(a) Model these data using a zero-inflated Poisson regression with grade responsible for structural zeros, and homework and gender predicting the counting portion. Write the model explicitly, estimating all parameters. Which predictors are significant at the 5% significance level?

(b) Is it a reliable model? Present the quantitative argument for the goodness-of-fit of the model.

(c) How are the estimated significant coefficients interpreted?

(d) What is the predicted number of books read by a second-grade girl for whom the reading is part of the homework?

EXERCISE 5.10. Thirty five patients in a large hospital were randomly chosen for a survey. The variables recorded were patient's BMI, age, gender, indicator of current smoking, and the number of mild to severe asthma attacks in the past three months. The data are summarized in the table below.

BMI	Age	Gender	Smoking	Attacks	BMI	Age	Gender	Smoking	Attacks
31.5	58	M	yes	0	33.7	71	F	no	1
25.6	32	M	no	2	23.9	65	F	no	1
31.2	26	M	no	2	28.6	50	M	yes	3
31.4	55	M	no	1	26.9	59	M	yes	2
33.5	75	F	no	0	23.8	60	M	no	0
33.9	53	M	no	0	31.3	62	M	yes	3
34.0	53	M	no	1	29.7	64	F	no	0
28.3	64	F	no	1	24.5	55	M	yes	5
23.1	53	F	no	0	25.1	31	F	yes	0
25.2	61	F	no	3	23.2	61	F	yes	2
20.6	61	F	yes	3	32.2	26	F	no	1
28.5	42	M	no	0	28.0	63	F	no	0
30.6	74	F	yes	2	28.1	70	F	no	1
27.4	72	M	yes	2	41.4	45	F	no	0
23.0	77	F	yes	4	24.8	86	F	yes	2
32.6	43	F	no	0	26.8	55	F	yes	3
25.8	74	F	yes	2	27.4	25	M	yes	3
37.9	72	M	yes	3					

(a) Run a ZIP model with smoking predicting the probability of excess zeros. Fit the model, estimate the parameters. Discuss significance of predictors.
(b) How good is the model fit?
(c) Interpret the estimates of the significant regression coefficients.
(d) Calculate the predicted value for the number of severe asthma attacks for a male patient, aged 60, whose BMI is 21.2, and who is currently a smoker.

EXERCISE 5.11. Consider the hurdle Poisson regression model defined by (5.12) – (5.14).
(a) Show that the expected value of the response variable has the form

$$\mathbb{E}\big(y|x_1, \ldots, x_k\big) = (1 - \pi) \frac{\lambda}{1 - \exp(-\lambda)}.$$

(b) Argue that the estimated regression coefficients in π and λ have the same interpretation as in a binary logistic and Poisson regression models, respectively.

EXERCISE 5.12. A coordinator of librarianship program within a school district is concerned with the negative effect of budget cuts on libraries. She randomly chooses 34 schools within the district and collects information on the number of computers, number of books (in thousands), number of journals during the current academic year, and the budget size (expenditure per student, in dollars). The data are:

Comps	Books	Jrnls	Budget	Comps	Books	Jrnls	Budget
0	8.2	0	0.00	22	9.0	16	0.00
19	11.7	10	16.45	32	18.3	23	22.22
0	2.0	0	5.29	0	12.0	5	0.17
13	8.2	8	23.50	6	8.8	12	7.14
5	30.0	2	6.33	1	14.0	60	1.83
16	14.1	15	7.20	5	12.5	32	24.66
12	9.5	0	3.07	7	3.0	5	7.07
6	21.8	0	4.00	3	16.3	40	12.00
12	9.0	11	4.39	1	6.5	40	13.85
22	5.0	20	17.07	3	8.5	4	18.22
0	15.7	4	1.82	4	10.0	20	30.49
7	19.3	66	9.09	7	18.0	100	0.81
6	20.8	2	10.49	0	11.5	2	0.61
28	11.0	30	0.47	3	9.1	0	9.19
0	9.3	0	0.06	13	10.4	36	25.67
11	12.7	14	0.00	36	7.5	55	7.89
17	15.6	14	22.22	0	19.7	8	1.00

(a) Run the hurdle Poisson regression to model the number of computers. Assume that if observations are positive, the number of computers is related to the number of books and periodicals, whereas the zero values are governed by expenditure per student. Write down the fitted model.

(b) Discuss the model fit.

(c) Interpret estimated significant parameters. State the practical conclusion.

(d) What is the predicted number of computers in a library with 10,000 books, 25 periodicals, and annual budget of $15 per student?

EXERCISE 5.13. Health care professionals conduct a study on medication adherence among senior citizens. They obtain medical and pharmaceutical records for a random sample of 30 patients (15 men, 15 women) who were prescribed the same once-daily heart medication. The variables that the investigators use for the analysis are: the number of days a patient forgot to take the prescribed medication (according to the pharmaceutical record), gender, age, and the number of other medications prescribed. The investigators suspect that older patients with more medications are more likely to account for zeros in the response variable, and that women have larger average positive responses than men. The collected data are given in the following table.

Days no meds	Gender	Age	Other meds	Days no meds	Gender	Age	Other meds
0	F	87	12	1	F	71	2
2	M	65	3	2	M	65	1
0	M	85	3	5	F	68	7
1	F	68	3	4	M	73	4
5	F	76	18	4	F	72	3
1	F	72	9	0	M	86	13
4	F	73	5	3	F	66	4
1	M	64	0	5	F	70	5
2	M	71	1	1	M	70	5
7	F	81	5	3	M	62	3
0	M	89	7	0	M	93	15
4	F	87	8	5	F	70	1
2	M	78	9	3	F	68	11
0	M	87	9	3	M	75	2
1	F	77	4	2	M	88	11

(a) Fit the hurdle Poisson model to verify the hypotheses. Identify all parameters in the predicted model. Is the conclusion supportive of the research hypotheses?

(b) How good is the model fit?

(c) Give interpretation of estimated significant regression coefficients.

(d) Predict the number of days with missed heart medication for a 78-year-old male patient who is prescribed to take only that one medication.

Chapter 6

Regression Models for Overdispersed Count Response

Suppose the response y is a count variable assuming non-negative integer values but unlike in the Poisson model, y may assume large values. In this chapter we consider four models that are reasonable alternatives to their Poisson-based models considered in the previous chapter: negative binomial, zero-truncated negative binomial, zero-inflated negative binomial, and hurdle negative binomial models.

6.1 Negative Binomial Regression Model

6.1.1 Model Definition

Recall that for a Poisson random variable, the mean is equal to the variance. A count variable for which the variance is larger than the mean is termed *overdispersed*. In this case a negative binomial regression would be a more appropriate model.

In a *negative binomial regression model*[1] with predictors $x_1, \ldots, x_k$, the response y follows a negative binomial distribution with the probability mass function given as:

$$\mathbb{P}(Y = y) = \left(\frac{r}{r + \lambda}\right)^r \frac{\Gamma(r + y)}{y! \, \Gamma(r)} \left(\frac{\lambda}{r + \lambda}\right)^y, \quad y = 0, 1, 2, \ldots, \quad (6.1)$$

---

[1] First rigorous treatment can be found in Hausman, J., Hall, B.H., and Z. Griliches (1984). "Econometric models for count data with an application to the patents – R & D relationship". *Econometrica*, 52(4): 909 – 938.

where

$$\lambda = \exp\left\{\beta_0 + \beta_1 x_1 + \cdots + \beta_k x_k\right\}, \quad \text{and } r \text{ is a positive constant.} \quad (6.2)$$

It can be proven (see Exercise 6.1) that $\mathbb{E}(y) = \lambda$ and $\mathbb{V}ar(y) = \lambda + \lambda^2/r$. Thus, if the response y has a large variance compared to its mean, the value of the parameter r is small. For this reason, r is called the *dispersion parameter*. If we let r go to infinity, the second term in the variance tends to zero, and, in fact, the limiting distribution is Poisson with rate λ.

Further, one can show (see Exercise 6.1) that the negative binomial distribution belongs to the exponential family of distributions, and thus the negative binomial model is an example of a generalized linear model with the log link function.

6.1.2 Fitted Model

From (6.2), in a fitted negative binomial regression model, the estimated parameters are $\widehat{\lambda} = \exp\left\{\widehat{\beta}_0 + \widehat{\beta}_1 x_1 + \cdots + \widehat{\beta}_k x_k\right\}$, and $\widehat{r}$.

6.1.3 Interpretation of Estimated Regression Coefficients

Since the mean of y is λ, the interpretation of estimated regression coefficients is identical to that in Poisson model (see Subsection 5.1.3).

6.1.4 Predicted Response

Considering the above expression for $\widehat{\lambda}$, for a given set of predictors $x_1^0, x_2^0, \ldots, x_k^0$, the predicted response y^0 is computed as $y^0 = \exp\left\{\widehat{\beta}_0 + \widehat{\beta}_1 x_1^0 + \cdots + \widehat{\beta}_k x_k^0\right\}$.

6.1.5 SAS Implementation

The `genmod` procedure invoked with the option `dist=negbin` fits a negative binomial regression.

• SAS outputs the quantity $1/r$ and terms it `dispersion parameter`. If the dispersion parameter is zero or close to zero, then the Poisson model is more appropriate. If it is much larger than zero, then negative binomial regression is valid.

6.1.6 R Implementation

In R, the function `glm.nb()` in the library `MASS` may be used to fit a negative binomial regression model. The general script looks like this:

`summary(`*fitted.model.name*`<- glm.nb(`*response.name* $\sim$ *x1.name* + $\cdots$ + *xk.name*`, data = `*data.name*`))`

- R outputs the estimate of the dispersion parameter r and calls it `Theta`.

6.1.7 Example

EXAMPLE 6.1. A college offers a 20-mile swim challenge program to all students, faculty, and staff. People who within 12 weeks complete a 20-mile swim in the pool on the college campus are awarded T-shirts. Number of laps swam is recorded in sets of 20 after every swim. The organizers have the data for 30 program participants for the first week of the program. They are interested in regressing the distance swam (in sets of 20 laps) on swimmer's gender, age, and whether it is their first time in this program. Since it is a self-paced program, it is expected to see a great variability in the number of sets swam during one week. The SAS code and relevant output are below.

```
data swim;
input gender$ age firsttime$ sets @@;
cards;
M 38 yes 20 M 26 no  0  M 21 yes 8
M 19 yes 13 M 18 yes 28 M 20 yes 2
F 26 yes 8  M 21 no  14 F 20 no  0
F 18 yes 3  M 25 yes 6  F 42 yes 1
F 24 yes 7  M 58 yes 27 M 19 yes 10
F 32 no  17 F 46 no  12 M 21 yes 4
F 26 no  3  M 22 no  35 F 19 yes 2
F 56 yes 11 F 41 no  15 M 25 no  1
M 25 yes 9  M 21 no  8  M 19 yes 11
M 37 no  34 F 22 yes 8  F 23 yes 5
;

proc genmod;
class gender(ref='F') firsttime(ref='yes');
 model sets=gender age firsttime/dist=negbin;
run;
```

Full Log Likelihood   -97.8206

Analysis Of Maximum Likelihood Parameter Estimates
Parameter   Estimate   Pr > ChiSq
Intercept   0.9316   0.0603
gender   M   0.8250   0.0082
age   0.0278   0.0474
firsttime no   0.2596   0.4021
Dispersion   0.5446

```
proc genmod;
model sets=/dist=negbin;
run;
```

Full Log Likelihood   -102.1982

```
data deviance_test;
  deviance=-2*(-102.1982-(-97.8206));
    pvalue=1-probchi(deviance,3);
run;

proc print;
run;
```

deviance   pvalue
 8.7552 0.032729

The estimates of the parameters in the model are $\widehat{\lambda} = \exp\{0.9316 + 0.8250 * male + 0.0278 * age + 0.2596 * not\,firsttime\}$ and $\widehat{r} = 1/0.5446 = 1.8362$. The model fits the data well since the p-value of the deviance test is below 0.05. The significant predictors are gender and age. The estimated average number of sets of laps swam by a male is $\exp\{0.8250\} \cdot 100\% = 228.19\%$ of that for female. For every one-year increase in age, the estimated average number of sets increases by $(\exp\{0.0278\} - 1) \cdot 100\% = 2.82\%$.

Suppose that we would like to predict the number of lap sets that a 20-year old female student who has never participated in the 20-mile swim challenge would swim during the first week. Calculated by hand, the predicted number of lap sets is $\exp\{0.9316 + 0.0278 * 20\} = 4.4264$. The same prediction is produced by SAS when running the following statements:

```
data prediction;
input gender$ age firsttime$;
cards;
F 20 yes
;

data swim;
 set swim prediction;
run;

proc genmod;
class gender(ref='F') firsttime(ref='yes');
 model sets=gender age firsttime/dist=negbin;
  output out=outdata p=pred_sets;
run;

proc print data=outdata(firstobs=31 obs=31);
var pred_sets;
run;

pred_sets
  4.42282
```

The R script and output for this example are below.

```
swim.data<-read.csv(file="./Example6.1Data.csv",
header=TRUE, sep=",")
install.packages("MASS")
library(MASS)

#specifying reference category
firsttime.rel<- relevel(swim.data$firsttime, ref="yes")

#fitting negative binomial model
summary(fitted.model <- glm.nb(sets ~ gender
+ age + firsttime.rel, data=swim.data))
```

Coefficients:

	Estimate	$Pr(>|z|)$
(Intercept)	0.93159	0.06034
genderM	0.82503	0.00784
age	0.02776	0.04222
firsttime.relno	0.25961	0.39579

Theta: 1.836

```
#checking model fit
intercept.only.model<- glm.nb(sets ~ 1, data=swim.data)
print(deviance<- -2*(logLik(intercept.only.model)
-logLik(fitted.model)))
```

8.755082

```
print(p.value<- pchisq(deviance, df=3, lower.tail=FALSE))
```

0.03273083

```
#using fitted model for prediction
print(predict(fitted.model, data.frame(gender="F", age=20,
firsttime.rel="yes"), type="response"))
```

4.422816

□

6.2 Zero-truncated Negative Binomial Regression Model

6.2.1 Model Definition

When a response variable y is a count variable that assumes only positive values and is overly dispersed, then a *zero-truncated negative binomial regression model*[2] maybe be appropriate. For predictors $x_1, \ldots, x_k$, the probability distribution function of the response variable y is modeled as a negative binomial truncated at zero:

$$\mathbb{P}(Y = y) = \frac{\left(\frac{r}{r+\lambda}\right)^r \frac{\Gamma(r+y)}{y!\,\Gamma(r)} \left(\frac{\lambda}{r+\lambda}\right)^y}{1 - \left(\frac{r}{r+\lambda}\right)^r}, \quad y = 1, 2, \ldots,$$

where $\lambda = \exp\{\beta_0 + \beta_1 x_1 + \cdots + \beta_k x_k\}$, and r is a positive constant. This model is not a generalized linear regression. Even though the distribution is a representative of the exponential family of distributions, the expected value of y is

$$\mathbb{E}(y) = \frac{\lambda}{1 - \left(\frac{r}{r+\lambda}\right)^r}, \tag{6.3}$$

thus, the log link function relates lambda but not the expected value of y to the linear regression.

---

[2]First considered in Gurmu, S. (1991). "Tests for detecting overdispersion in the positive Poisson regression model". *Journal of Business and Economic Statistics*, 9(2): 215 – 222.

6.2.2 Fitted Model

In a fitted zero-truncated negative binomial regression model, the parameter estimates are $\widehat{\lambda} = \exp\left\{\widehat{\beta}_0 + \widehat{\beta}_1 x_1 + \cdots + \widehat{\beta}_k x_k\right\}$, and $\widehat{r}$.

6.2.3 Interpretation of Estimated Regression Coefficients

Since the mean of y, given in (6.3), describes a relatively complex relation between the response and predictors, in this model there is no easy interpretation of the estimated regression coefficients. Traditionally, interpretation is omitted.

6.2.4 Predicted Response

Taking into account (6.3) and the above expression for the estimate of λ, for a given set of predictors $x_1^0, x_2^0, \ldots, x_k^0$, the predicted response is computed as

$$y^0 = \frac{\widehat{\lambda}}{1 - \left(\frac{\widehat{r}}{\widehat{r}+\widehat{\lambda}}\right)^{\widehat{r}}} = \frac{\widehat{\lambda}}{1 - \left(1 + \widehat{\lambda}/\widehat{r}\right)^{-\widehat{r}}}$$

$$= \frac{\exp\left\{\widehat{\beta}_0 + \widehat{\beta}_1 x_1^0 + \cdots + \widehat{\beta}_k x_k^0\right\}}{1 - \left(1 + \exp\left\{\widehat{\beta}_0 + \widehat{\beta}_1 x_1^0 + \cdots + \widehat{\beta}_k x_k^0\right\}/\widehat{r}\right)^{-\widehat{r}}}.$$

6.2.5 SAS Implementation

A zero-truncated negative binomial model may be fitted by applying the `fmm` procedure with the option `dist=truncnegbin` in the model statement.

• SAS produces the estimate $1/\widehat{r}$, the reciprocal of the estimated dispersion parameter. The term that SAS uses for its estimate is `Scale Parameter`.

6.2.6 R Implementation

The function `vglm()` in the library `VGAM` may be used to fit a zero-truncated negative binomial regression model in R. The script is

```
summary(fitted.model.name<- vglm(response.name ~ x1.name + ···
+ xk.name, data = data.name, family = posnegbinomial()))
```

As part of the output, two intercepts are fitted, called (Intercept):1 and (Intercept):2. The first one is the regression intercept $\widehat{\beta}_0$, whereas the second one is $\ln(\widehat{r})$, the natural logarithm of the estimated dispersion parameter.

6.2.7 Example

EXAMPLE 6.2. An assistant general manager of a luxury hotel is put in charge of implementing a budget-friendly new ideas to help reduce the number of complaints by hotel guests. She collects data on a random sample of 28 guests who made at least one complaint, recording the number of complaints made by the guest, room floor, whether the guest is an elite member, and the duration of the stay (in days). SAS code and output are given below.

```
data hotel;
input   ncomplaints floor member$ days@@;
cards;
2  6 no  3   1 7 no  12   3 6 no   7   3 7 yes 3
3  8 yes 2   9 3 no  4    1 8 no   4   2 6 yes 5
6  5 no  8   2 8 no  11   1 2 yes  5   2 3 no  3
2  4 no  8   2 3 no  6    7 4 yes  4   4 5 yes 5
8  2 yes 3   3 5 no  8    4 5 yes  4   4 2 no  3
1  2 yes 4   1 7 no  3    6 3 no   2   3 3 no  2
12 2 yes 4   2 6 no  1    1 6 no   3   5 3 no  8
;

proc format;
value $memberfmt 'no'='ref' 'yes'='member';
run;

proc fmm;
class member;
 model ncomplaints=floor member days/dist=truncnegbin;
format member $memberfmt.;
run;
```

-2 Log Likelihood 108.6

Parameter Estimates for Truncated Negative Binomial Model

Effect	Estimate	Pr > |z|
Intercept	1.9919	<.0001

floor	-0.2400	0.0076
member member	0.3712	0.2560
member ref	0	.
days	0.006069	0.9319
Scale Parameter	0.2446	

```
proc fmm;
  model ncomplaints=/dist=trucnegbin;
run;
```

-2 Log Likelihood 118.1

```
data deviance_test;
 deviance=118.1-108.6;
   pvalue=1-probchi(deviance,3);
run;
```

```
proc print;
run;
```

deviance	pvalue
9.5	0.023331

The fitted model has parameter estimates $\widehat{\lambda} = \exp\{1.9919 - 0.2400 * floor + 0.3712 * member + 0.006069 * days\}$, and $\widehat{r} = 1/0.2446 = 4.0883$. The model has a decent fit since the p-value in the deviance test is smaller than 0.05. In this model, the room floor is the only significant predictor at the 5% level. As mentioned above, the estimated coefficients do not yield a straightforward interpretation.

To predict the number of complaints made by an elite member who stays for two nights and whose room is on the forth floor, we compute:

$$y^0 = \frac{\exp\{1.9919 - 0.2400 * 4 + 0.3712 + 0.006069 * 2\}}{1 - \left(1 + \exp\{1.9919 - 0.2400 * 4 + 0.3712 + 0.006069 * 2\}/4.0883\right)^{-4.0883}}$$
$$= 4.3707.$$

This is the same prediction as in SAS, which can be seen by running this code:

```
data prediction;
input floor member$ days;
cards;
4 yes 2
```

```
;

data hotel;
 set hotel prediction;
run;

proc fmm;
class member;
 model ncomplaints=floor member days/dist=truncnegbin;
 output out=outdata pred=p_ncomplaints;
run;

proc print data=outdata(firstobs=29 obs=29);
var p_ncomplaints;
run;
```

p_ncomplaints
    4.37130

The R code that reproduces the output in SAS is as follows:

```
hotel.data<-read.csv(file="./Example6.2Data.csv",
header=TRUE, sep=",")
install.packages("VGAM")
library(VGAM)

#fitting truncated negative binomial model
summary(fitted.model<- vglm(ncomplaints ~ floor + member +
days, data=hotel.data, family=posnegbinomial()))
```

Coefficients:

	Estimate	Pr(>|z|)
(Intercept):1	1.991913	6.79e-05
(Intercept):2	1.408035	0.08584
floor	-0.239965	0.00523
memberyes	0.371197	0.24786
days	0.006074	0.93034

```
#checking model fit
intercept.only.model<- vglm(ncomplaints ~ 1, data=hotel.data,
family=posnegbinomial())
print(deviance<- -2*(logLik(intercept.only.model)
-logLik(fitted.model)))
```

9.489394

```
print(p.value<- pchisq(deviance, df=3, lower.tail=FALSE))
```

0.02344446

```
#using fitted model for prediction
print(predict(fitted.model, data.frame(floor=4, member="yes",
days=2), type="response"))
```

4.371325

Note that since `(Intercept):2 1.408035`, the estimated dispersion parameter is $\widehat{r} = \exp\{1.408035\} = 4.0879$. $\square$

6.3 Zero-inflated Negative Binomial Regression Model

6.3.1 Model Definition

A *zero-inflated negative binomial (ZINB) regression model*[3] is used when the response is a count variable that exhibits overdispersion and has a large number of zeros which cannot be explained through a chance alone. The ZINB model better accounts for overdispersed characteristics than the zero-inflated Poisson regression. In the ZINB regression model, the response variable has the probability mass function defined by:

$$\mathbb{P}(Y = y) = \begin{cases} \pi + (1 - \pi)\left(\dfrac{r}{r + \lambda}\right)^r, & \text{if } y = 0, \\ (1 - \pi)\left(\dfrac{r}{r + \lambda}\right)^r \dfrac{\Gamma(r + y)}{y!\,\Gamma(r)}\left(\dfrac{\lambda}{r + \lambda}\right)^y, & \text{if } y = 1, 2, \ldots, \end{cases}$$

where

$$\pi = \frac{\exp\{\beta_0 + \beta_1 x_1 + \cdots + \beta_m x_m\}}{1 + \exp\{\beta_0 + \beta_1 x_1 + \cdots + \beta_m x_m\}}, \tag{6.4}$$

$$\lambda = \exp\{\gamma_0 + \gamma_1 x_{m+1} + \cdots + \gamma_{k-m} x_k\}, \quad \text{and } r \text{ is a positive constant.} \tag{6.5}$$

The zero-inflated negative binomial distribution is a mixture of two distributions and, thus, is not a member of the exponential family of distributions. Consequently, the model doesn't belong to the class of generalized linear models.

---

[3]Originally considered in Greene, W. H. (1994). "Some Accounting for Excess Zeros and Sample Selection in Poisson and Negative Binomial Regression Models". *Working Paper EC-94-10: Department of Economics, New York University.*

147

6.3.2 Fitted Model

By (6.4) and (6.5), the fitted zero-inflated negative binomial model has the estimated parameters of the form:

$$\widehat{\pi} = \frac{\exp\{\widehat{\beta}_0 + \widehat{\beta}_1 x_1 + \cdots + \widehat{\beta}_m x_m\}}{1 + \exp\{\widehat{\beta}_0 + \widehat{\beta}_1 x_1 + \cdots + \widehat{\beta}_m x_m\}}, \tag{6.6}$$

$$\widehat{\lambda} = \exp\{\widehat{\gamma}_0 + \widehat{\gamma}_1 x_{m+1} + \cdots + \widehat{\gamma}_{k-m} x_k\}, \quad \text{and } \widehat{r}. \tag{6.7}$$

6.3.3 Interpretation of Estimated Regression Coefficients

The mean of the response is (show it!) $\mathbb{E}(y) = (1 - \pi)\lambda$, thus the interpretation of the estimated regression coefficients coincides with that in the ZIP model (see Subsection 5.3.3), that is, the estimated beta coefficients are interpreted as in a binary logistic model, whereas the estimated gamma coefficients are interpreted as in the Poisson model.

6.3.4 Predicted Response

Making use of the formula for the mean response $\mathbb{E}(y) = (1 - \pi)\lambda$, and expressions (6.6) and (6.7), we see that for some concrete values of predictors $x_1^0, \ldots, x_k^0$, the predicted response is calculated as:

$$y^0 = \left(1 - \frac{\exp\{\widehat{\beta}_0 + \widehat{\beta}_1 x_1^0 + \cdots + \widehat{\beta}_m x_m^0\}}{1 + \exp\{\widehat{\beta}_0 + \widehat{\beta}_1 x_1^0 + \cdots + \widehat{\beta}_m x_m^0\}}\right) \exp\{\widehat{\gamma}_0 + \widehat{\gamma}_1 x_{m+1}^0 + \cdots + \widehat{\gamma}_{k-m} x_k^0\}$$

$$= \frac{\exp\{\widehat{\gamma}_0 + \widehat{\gamma}_1 x_{m+1}^0 + \cdots + \widehat{\gamma}_{k-m} x_k^0\}}{1 + \exp\{\widehat{\beta}_0 + \widehat{\beta}_1 x_1^0 + \cdots + \widehat{\beta}_m x_m^0\}}.$$

6.3.5 SAS Implementation

The genmod procedure with the option dist=zinb and zeromodel statement fits a zero-inflated negative binomial regression. The syntax is:

```
proc genmod data=data_name;
    class <list of categorical predictors>;
        model response_name=<list of predictors>/dist=zinb;
            zeromodel <list of predictors of structural zeros>;
run;
```

- SAS outputs the estimate of the dispersion parameter (called `Dispersion`) that is equal to $1/\hat{r}$.

6.3.6 R Implementation

The function `zeroinfl()` in the library `pscl` may be used to fit a zero-inflated negative binomial regression in R. The syntax for this function is

summary(*fitted.model.name*<- `zeroinfl`(*response.name* ~ *x{m+1}.name*
+ ··· + *xk.name|x1.name* + ··· + *xm.name*, `data` = *data.name*,
`dist = "negbin")`

- R prints the estimate of the dispersion parameter r and terms it `Theta`.

6.3.7 Example

EXAMPLE 6.3. Sixty Californian weather stations with continuous records between 1965 and 2014 were randomly chosen for the analysis. For each station, its elevation (in meters), overall minimum temperature (in Fahrenheit), and the number of years when it snowed were considered. The number of snowy years ranged between 0 and 50, with 26 stations having zero years of snow. We fit the zero-inflated negative binomial regression with the elevation as the predictor of π, and the overall minimum temperature as the predictor of λ. The SAS code is presented below.

```
data weather;
 input elevation mintemp snowyears @@;
cards;
131.1   72.6 0  81.7     68.7 0  602      65.4 2
1338.1 58.8 50 18.3     67.0 2  310.3  64.6 1
18.3    67.5 2  182.9    67.9 0  1444.4 54.5 44
1278.6 57.0 27 1389.9   57.7 49 1295.4 59.9 21
712.6   62.2 7  256      69.2 0  27.4    65.4 0
974.8   56.3 44 271.3    70.7 0  381.9  63.9 26
71       68.9 0  1088.1   66.8 12 3.7      70.9 0
28       73.1 0  520.6    65.4 6  2139.7 60.1 24
974.8   59.1 44 18.3     67.6 0  18       67.7 0
101.5   70.3 2  44.2     68.2 0  327.7  68.0 6
125      68.0 0  999.7    62.1 50 146.3  67.4 0
77.4    66.6 1  146.3    68.7 0  240.8  69.9 0
391.7   65.4 0  222.5    66.9 0  1165.9 63.1 22
132.6   65.8 0  483.4    65.0 1  712.6  64.7 7
```

```
847.6   61.7 45 725.4    66.5 16 36.6     65.9 0
81.7    68.6 0  6.1      69.1 0  73.8     67.4 0
1516.4  55.5 50 1160.7   61.5 50 111.3    67.9 1
21      67.6 1  12.2     66.2 0  83.8     66.2 0
1160.7  62.5 50 576.4    67.2 0  398.4    64.2 4
1431    59.6 41 41.5     67.9 7  1179.6   64.0 49
;

proc genmod;
 model snowyears=mintemp/dist=zinb;
  zeromodel elevation;
run;
```

Full Log Likelihood   -151.5915

Analysis Of Maximum Likelihood Parameter Estimates

Parameter	Estimate	Pr > ChiSq
Intercept	19.0986	<.0001
mintemp	-0.2593	<.0001
Dispersion	0.7447	

Analysis Of Maximum Likelihood Zero Inflation Parameter Estimates

Parameter	Estimate	Pr > ChiSq
Intercept	1.2623	0.0228
elevation	-0.0056	0.0054

```
proc genmod;
 model snowyears=/dist=zinb;
  zeromodel;
run;
```

Full Log Likelihood   -178.2204

```
data deviance_test;
 deviance=-2*(-178.2204-(-151.5915));
   pvalue=1-probchi(deviance,2);
run;

proc print;
run;
```

deviance	pvalue
53.2578	2.724E-12

The fitted model has parameter estimates $\widehat{\pi} = \exp\{1.2623-0.0056*elevation\}/$ $(1+\exp\{1.2623-0.0056*elevation\})$, $\widehat{\lambda} = \exp\{19.0986-0.2593*mintemp\}$, and $\widehat{r} = 1/0.7447 = 1.3428$. The model has a good fit as judged by the tiny p-value for the goodness-of-fit test. Both minimum temperature and elevation are significant predictors in this model. For a one-meter increase in elevation, the estimated odds in favor of zero years with snow change by $(\exp\{-0.0056\} - 1) \cdot 100\% = -0.56\%$, that is, decrease by 0.56%. For a one-degree increase in minimum temperature, the estimated average number of snowy years changes by $(\exp\{-0.2593\} - 1) \cdot 100\% = -22.84\%$, or decreases by 22.84%.

Consider the station at the elevation of 1,165.9 meters where the minimum temperature is 63.1 degrees Fahrenheit. The observed number of years when it snowed is 22. The value predicted by the model is found as

$$y^0 = \frac{\exp\{19.0986 - 0.2593 * 63.1\}}{1 + \exp\{1.2623 - 0.0056 * 1165.9\}} = 15.3578.$$

The same predicted value is outputted by SAS if we run these statements:

```
data prediction;
input elevation mintemp;
cards;
1165.9 63.1
;

data weather;
 set weather prediction;
run;

proc genmod;
 model snowyears=mintemp/dist=zinb;
  zeromodel elevation;
   output out=outdata p=p_snowyears;
run;

proc print data=outdata(firstobs=61 obs=61);
var p_snowyears;
run;

p_snowyears
    15.3545
```

The R code that produces matching output is below.

```
weather.data<-read.csv(file="./Example6.3Data.csv",
header=TRUE, sep=",")
install.packages("pscl")
library(pscl)

#fitting zero-inflated negative binomial model
summary(fitted.model<- zeroinfl(snowyears ~ mintemp|elevation,
data = weather.data, dist = "negbin"))
```

Count model coefficients (negbin with log link):

	Estimate	Pr(>|z|)
(Intercept)	19.09858	4.04e-09
mintemp	-0.25931	5.23e-07

Zero-inflation model coefficients (binomial with logit link):

	Estimate	Pr(>|z|)
(Intercept)	1.262349	0.02124
elevation	-0.005642	0.00349

Theta = 1.3428

```
#checking model fit
intercept.only.model<- zeroinfl(snowyears ~ 1,
  data=weather.data,
dist="negbin")
print(deviance<- -2*(logLik(intercept.only.model)
-logLik(fitted.model)))
```

53.25778

```
print(p.value<- pchisq(deviance, df=2, lower.tail=FALSE))
```

2.724081e-12

```
#using fitted model for prediction
print(predict(fitted.model,
data.frame(elevation=1165.9, mintemp=63.1)))
```

15.35455

□

6.4 Hurdle Negative Binomial Regression Model

6.4.1 Model Definition

For a count response variable with overdispersion, zeros may be modeled independently of positive responses. The *hurdle negative binomial regression model*[4] accomplishes just that. In this model, assuming $x_1, \ldots, x_k$ are the predictors, the response variable y has the probability distribution function

$$
\mathbb{P}(Y = y) = \begin{cases} \pi, & \text{if } y = 0, \\ (1 - \pi) \dfrac{\left(\frac{r}{r+\lambda}\right)^r \frac{\Gamma(r+y)}{y!\,\Gamma(r)} \left(\frac{\lambda}{r+\lambda}\right)^y}{1 - \left(\frac{r}{r+\lambda}\right)^r}, & \text{if } y = 1, 2, \ldots, \end{cases}
$$

where

$$
\pi = \frac{\exp\{\beta_0 + \beta_1 x_1 + \cdots + \beta_m x_m\}}{1 + \exp\{\beta_0 + \beta_1 x_1 + \cdots + \beta_m x_m\}}, \tag{6.8}
$$

$\lambda = \exp\left\{\gamma_0 + \gamma_1 x_{m+1} + \cdots + \gamma_{k-m} x_k\right\}$, and r is a positive constant. (6.9)

This model is not a generalized linear model, since the distribution is a mixture distribution.

6.4.2 Fitted Model

From (6.8) and (6.9), in the fitted hurdle negative binomial regression model, the estimated parameters take the form

$$
\widehat{\pi} = \frac{\exp\{\widehat{\beta}_0 + \widehat{\beta}_1 x_1 + \cdots + \widehat{\beta}_m x_m\}}{1 + \exp\{\widehat{\beta}_0 + \widehat{\beta}_1 x_1 + \cdots + \widehat{\beta}_m x_m\}}, \tag{6.10}
$$

$$
\widehat{\lambda} = \exp\left\{\widehat{\gamma}_0 + \widehat{\gamma}_1 x_{m+1} + \cdots + \widehat{\gamma}_{k-m} x_k\right\}, \quad \text{and} \quad \widehat{r}. \tag{6.11}
$$

6.4.3 Interpretation of Estimated Regression Coefficients

The estimated beta coefficients are interpreted as in the logistic regression, in terms of the odds in favor of zero. Also, it can be shown (do it!) that the expected response depends on the parameters via the relation

$$
\mathbb{E}(y) = (1 - \pi) \frac{\lambda}{1 - \left(\frac{r}{r+\lambda}\right)^r} = (1 - \pi) \frac{\lambda}{1 - \left(1 + \lambda/r\right)^{-r}}.
$$

---

[4]First studied in Mullahy, J. (1986). "Specification and testing of some modified count data models". *Journal of Econometrics*, 33(3): 341 – 365.

This relation is too complex to yield an easy interpretation of the estimated gamma regression coefficients.

6.4.4 Predicted Response

In view of the above expression for the expected value of y, and also (6.10) and (6.11), the predicted value y^0, when predictor variables assume certain fixed values $x_1^0, \ldots, x_k^0$, can be found as

$$
y^0 = \left(1 - \frac{\exp\{\widehat{\beta}_0 + \widehat{\beta}_1 x_1^0 + \cdots + \widehat{\beta}_m x_m^0\}}{1 + \exp\{\widehat{\beta}_0 + \widehat{\beta}_1 x_1^0 + \cdots + \widehat{\beta}_m x_m^0\}} \right) \times
$$

$$
\times \frac{\exp\left\{\widehat{\gamma}_0 + \widehat{\gamma}_1 x_{m+1}^0 + \cdots + \widehat{\gamma}_{k-m} x_k^0\right\}}{1 - \left(1 + \exp\left\{\widehat{\gamma}_0 + \widehat{\gamma}_1 x_{m+1}^0 + \cdots + \widehat{\gamma}_{k-m} x_k^0\right\}/\widehat{r} \right)^{-\widehat{r}}}
$$

$$
= \frac{\left(1 + \exp\{\widehat{\beta}_0 + \widehat{\beta}_1 x_1^0 + \cdots + \widehat{\beta}_m x_m^0\} \right)^{-1} \exp\left\{\widehat{\gamma}_0 + \widehat{\gamma}_1 x_{m+1}^0 + \cdots + \widehat{\gamma}_{k-m} x_k^0\right\}}{1 - \left(1 + \exp\left\{\widehat{\gamma}_0 + \widehat{\gamma}_1 x_{m+1}^0 + \cdots + \widehat{\gamma}_{k-m} x_k^0\right\}/\widehat{r} \right)^{-\widehat{r}}}.
$$

6.4.5 SAS Implementation

Invoking the `fmm` procedure with the `dist=truncnegbin` option in the `model` statement fits the hurdle negative binomial model. The complete syntax follows.

```
proc fmm data=data_name;
    class <list of categorical predictors>;
      model response_name=<list of predictors>/dist=truncnegbin;
       model response_name=/dist=constant;
        probmodel <list of predictors of zeros>;
     output out=outdata_name pred;
run;
```

• As explained in Subsection 5.4.5, when writing down the fitted model, the estimated regression coefficients in $\widehat{\pi}$ have to be taken with the opposite sign.
• SAS estimates the inverse of the dispersion parameter r, and terms it `Scale Parameter`.

6.4.6 R Implementation

The function `hurdle()` in the library `pscl` may be used to fit a hurdle negative binomial model in R. The syntax in this case is

summary($fitted.model.name$<- `hurdle`($response.name \sim x\{m+1\}.name + \cdots$
$+ xk.name \mid x1.name + \cdots + xm.name$, `data=`$data.name$, `dist="negbin"`,
`zero.dist= "binomial"`, `link="logit"`))

- R prints the estimate of the dispersion parameter. The name for the quantity is `Theta`.

6.4.7 Example

EXAMPLE 6.4. A bank wants to estimate the risk of delinquent credit card accounts for new applicants. A random sample of 35 applicants is drawn. The selected variables are the number of previously delinquent credit card accounts, age, gender, current income (high/low), and the total number of years ever unemployed. The conjecture is that age could account for zero delinquent accounts, whereas gender, income, and the number of unemployed years might be associated with positive responses. The code that fits a hurdle negative binomial model is as follows:

```
data creditcards;
input ndelinqaccounts age gender$ income$ nunemplyears @@;
cards;
12 53 M Low  8    0 26 F Low  4   16 49 M Low  8    0 23 M Low  5
7  28 F High 6    2 44 M Low  2   11 49 F Low  7    4 43 M Low  4
0  25 M Low  0    0 28 M High 5    4 25 M High 1    5 40 M High 6
1  37 M Low  3    0 48 M High 1    7 36 F Low  5    0 48 M Low  4
13 56 M Low  10   0 22 M Low  0    7 36 F High 4    3 35 M Low  1
7  35 F High 0    0 42 F Low  0    1 56 M Low  4    5 22 M Low  0
0  38 F Low  7    4 52 M High 5    1 30 F Low  6    0 27 M High 1
0  32 F High 3    4 46 F Low  1    2 32 M Low  2    7 26 F Low  6
0  23 M Low  5    3 30 M High 2    0 25 M Low  3
;

proc fmm;
class gender income;
  model ndelinqaccounts=gender income
    nunemplyears/dist=truncnegbin;
```

```
model+/dist=constant;
  probmodel age;
run;
```

-2 Log Likelihood 144.5

Parameter Estimates for Truncated Negative Binomial Model

Effect	gender	income	Estimate	Pr > |z|
Intercept			0.7462	0.0153
gender	F		0.2343	0.3352
gender	M		0	.
income		High	0.1526	0.5687
income		Low	0	.
nunemplyears			0.1740	<.0001
Scale Parameter			0.09103	

Parameter Estimates for Mixing Probabilities

Effect	Estimate	Pr > |z|
Intercept	-2.3203	0.1006
age	0.08147	0.0442

```
proc fmm;
  model ndelinqaccounts=/dist=truncnegbin;
   model+/dist=constant;
    probmodel;
run;
```

-2 Log Likelihood 161.4

```
data deviance_test;
  deviance=161.4-144.5;
   pvalue=1-probchi(deviance,4);
run;

proc print;
run;
```

deviance	pvalue
16.9	0.002021359

In the fitted model, the estimated parameters are $\widehat{\pi} = \frac{\exp\{2.3203-0.08147*age\}}{1+\exp\{2.3203-0.08147*age\}}$, $\widehat{\lambda} = \exp\{0.7462 + 0.2343 * female + 0.1526 * high\_income + 0.1740 * nunemplyears\}$, and $\widehat{r} = 1/0.09103 = 10.9854$. This model fits the data well,

156

as demonstrated by a small p-value in the deviance test. Age is a significant predictor of π, whereas the number of unemployed years is the only significant predictor of λ. For a one-year increase in age, the estimated odds in favor of zero delinquent accounts changes by $\exp\{-0.08147\} - 1) \cdot 100\% = -7.82\%$, or is reduced by 7.82%. Further, due to complexity of the model, we will not attempt to interpret the estimated gamma regression coefficients.

To predict the number of delinquent accounts for a 45-year old male with high income and zero years of unemployment, we compute

$$y^0 = \frac{\exp\{0.7462 + 0.1526\}\left(1 + \exp\{2.3203 - 0.08147 * 45\}\right)^{-1}}{1 - \left(1 + \exp\{0.7462 + 0.1526\}/10.9854\right)^{-10.9854}} = 2.1878.$$

The same value is outputted in SAS:

```
data prediction;
input age gender$ income$ nunemplyears;
cards;
45 M High 0
;

data creditcards;
 set creditcards prediction;
run;

proc fmm;
class gender income;
  model ndelinqaccounts=gender income
    nunemplyears/dist=truncnegbin;
  model +/dist=constant;
   probmodel age;
    output out=outdata pred=p_ndelinqaccounts;
run;

proc print data=outdata(firstobs=36 obs=36);
var p_ndelinqaccounts;
run;

p_ndelinqaccounts
        2.18752
```

The script and output in R are given below.

```
creditcards.data<-read.csv(file="./Example6.4Data.csv",
header=TRUE, sep=",")
install.packages("pscl")
library(pscl)

#specifying reference categories
gender.rel<- relevel(creditcards.data$gender, ref="M")
income.rel<- relevel(creditcards.data$income, ref="Low")

#fitting hurdle negative binomial model
summary(fitted.model<- hurdle(ndelinqaccounts ~ gender.rel
+ income.rel + nunemplyears|age, data=creditcards.data,
dist="negbin", zero.dist="binomial", link="logit"))
```

Count model coefficients (truncated negbin with log link):

	Estimate	Pr(>|z|)
(Intercept)	0.74619	0.0153
gender.relF	0.23427	0.3352
income.relHigh	0.15257	0.5687
nunemplyears	0.17401	7.51e-05

Zero hurdle model coefficients (binomial with logit link):

	Estimate	Pr(>|z|)
(Intercept)	-2.32034	0.1006
age	0.08147	0.0442

Theta: count = 10.9851

```
#checking model fit
intercept.only.model<- hurdle(ndelinqaccounts ~ 1,
data=creditcards.data, dist="negbin",
zero.dist="binomial", link="logit")
print(deviance<- -2*(logLik(intercept.only.model)
-logLik(fitted.model)))
```

16.80772

```
print(p.value<- pchisq(deviance, df=4, lower.tail=FALSE))
```

0.002106471

```
#using fitted model for prediction
print(predict(fitted.model, data.frame(age=45, gender.rel="M",
income.rel="High", nunemplyears=0)))
```

2.187517

□

Exercises for Chapter 6

EXERCISE 6.1. Consider a random experiment consisting of a sequence of independent trials each with outcomes of success or failure. And let p denote the probability of a success.

(a) Argue that X, the number of successes observed until the rth failure, is a negative binomial random variable with the probability mass function

$$P(X = x) = \binom{x + r - 1}{x} p^x (1 - p)^r, \quad x = 0, 1, 2, \ldots.$$

Show that for a fixed r, this distribution is a special case of the exponential family of distributions, that is, it can be written in the form (1.3) where $\theta = \ln p$ and $\phi = 1$.

(b) Using the substitution $p = \dfrac{\lambda}{r + \lambda}$, rewrite the probability mass function in part (a) as

$$P(X = x) = \left(\frac{r}{r + \lambda}\right)^r \frac{\Gamma(r + x)}{x! \, \Gamma(r)} \left(\frac{\lambda}{r + \lambda}\right)^x, \quad x = 0, 1, 2, \ldots.$$

Derive that $\mathbb{E}(X) = \lambda$ and $\mathbb{V}ar(X) = \lambda + \lambda^2/r$. Hint: use the fact that for a negative binomial random variable with a mass function defined in part (a), the mean is $\dfrac{pr}{1 - p}$, and variance is $\dfrac{pr}{(1 - p)^2}$.

(c) Suppose the parameter r goes to infinity, but the mean λ is constant. Show that the limiting probability mass function is that of a Poisson distribution, that is, prove that

$$\lim_{r \to \infty} \left(\frac{r}{r + \lambda}\right)^r \frac{\Gamma(r + x)}{x! \, \Gamma(r)} \left(\frac{\lambda}{r + \lambda}\right)^x$$

$$= \frac{\lambda^x}{x!} \lim_{r \to \infty} \frac{\Gamma(r + x)}{\Gamma(r) \, (r + \lambda)^x} \left(1 + \frac{\lambda}{r}\right)^{-r} = \frac{\lambda^x}{x!} e^{-\lambda}, \quad x = 0, 1, 2, \ldots.$$

EXERCISE 6.2. Graduate students in marine biology conducted an experiment on mussel *Mytilus californicus* survival. They prepared 24 cages with 50 mussels each, and placed them in various plots in tidal areas along the shoreline. The food was supplied daily at high, medium, or low levels. Maximum and minimum daily temperatures were logged by an automatic data recorder positioned in each cage. The number of mussels that died in each cage was recorded two weeks later. The data are:

Max temp	Min temp	Feeding level	N dead mussels	Max temp	Min temp	Feeding level	N dead mussels
77	60	high	0	83	62	med	8
88	59	high	1	75	63	med	3
78	62	high	1	76	61	med	2
85	60	high	2	86	62	med	1
78	61	high	0	92	62	low	2
89	63	high	0	89	64	low	3
92	62	high	2	96	68	low	19
75	58	high	0	86	62	low	7
80	59	med	1	74	61	low	3
90	61	med	2	88	62	low	12
74	63	med	4	97	63	low	9
92	62	med	6	91	61	low	7

(a) Model mussel mortality via the negative binomial regression. Present the fitted model. What predictors turn out to be significant at the 5% level?
(b) How good it the model fit?
(c) How would you interpret the estimated significant coefficients?
(d) Predict the number of dead mussels that were fed a hight level of food, and were located in an area with the maximum temperature of 75 degrees and minimum temperature of 60 degrees.

EXERCISE 6.3. Researchers in K-12 education are conducting a study on daily allowance that high-school students receive from their parents. They randomly select 30 students and record their age, gender, whether they have had a job during the past summer, and an approximate daily allowance (in units of $5). The data are:

Age	Gender	Job	Allowance	Age	Gender	Job	Allowance
15	M	yes	0	15	M	no	0
18	F	yes	3	15	M	no	12
18	M	yes	3	18	M	no	3
14	F	no	6	15	M	no	4
16	F	yes	2	18	M	yes	0
17	F	yes	1	15	F	no	8
18	F	yes	1	15	M	no	5
15	F	no	4	15	M	no	5
16	M	yes	1	14	M	no	4
16	F	no	9	16	F	yes	3
16	M	no	3	17	M	no	2
16	M	no	10	18	M	yes	2
16	F	yes	0	17	F	yes	11
14	M	no	9	15	M	no	6
17	M	yes	1	16	M	no	12

(a) Is the negative binomial regression appropriate in modeling the amount of daily allowance? Fit the model and discuss significance of the predictor variables.

(b) How good is the model fit?

(c) Interpret the estimated significant regression coefficients.

(d) Predict the amount of daily allowance for a male student, age 16, who hasn't held a summer job.

EXERCISE 6.4. A state park is interested in proper allocation of recreational resources. As part of the analysis, data were collected on 30 randomly chosen kayak rentals. The number of kayaks rented, party size, length of route (in hours), and whether the party camped overnight (and returned kayaks the next day). Note that some parties owned some number of kayaks and rented just a few, so party size is not necessarily linearly proportional to the number of rented kayaks. In addition, rented kayaks could be two- or three-seaters. The data are given below.

N rented kayaks	Party size	Route length	Camped overnight	N rented kayaks	Party size	Route length	Camped overnight
6	12	1	yes	7	14	3	no
2	4	3	yes	2	7	12	no
3	7	12	no	2	6	12	yes
2	6	3	no	3	18	6	yes
1	3	2	no	2	4	1	yes
2	7	6	yes	2	4	4	yes
2	4	2	no	4	9	12	yes
1	3	6	yes	3	10	2	no
3	9	12	yes	1	2	3	no
5	10	4	no	6	12	4	no
1	2	1	no	5	12	12	yes
2	6	12	no	2	7	6	yes
4	9	4	no	3	8	12	no
1	3	2	no	7	14	3	no
3	7	2	no	1	3	6	yes

(a) Argue that a zero-truncated negative binomial regression would be appropriate to model the number of rented kayaks. Fit the model. Discuss significance of predictors.

(b) Discuss model fit.

(c) Interpret the estimated significant regression coefficients, whatever are possible to interpret.

(d) From the given data, determine an "average" party, the one that has the average party size, the average route length, and the most frequent observation for camping overnight, and predict the number of rented kayaks for this "average" party.

EXERCISE 6.5. A sociologist is studying popularity of YouTube vlogs. He selects a random sample of 40 video channels and records the number of new videos posted in the past month, total number of videos posted by the vlogger, number of subscribers (in thousands), number of views (in thousands), and type of videos (comedy, fashion, life advice, news, popular science, or new products and brands). The data are summarized in the following table.

N new videos	Num videos	Num subscr	Num views	Video type	N new videos	Num videos	Num subscr	Num views	Video type
3	81	3.9	205.8	life advice	3	212	1.3	121.1	products
4	188	27	213.6	fashion	4	86	4.2	160.2	fashion
1	55	10.1	176.8	products	12	517	85.4	163.7	life advice
4	123	14.4	59.7	science	7	100	8	91.5	news
1	65	5	508.7	life advice	9	130	2.8	38.9	life advice
2	118	3.5	280.6	comedy	2	34	2.4	151.9	fashion
3	119	4.7	25.7	fashion	30	396	7.6	118.4	comedy
1	47	4.4	135.8	products	12	52	0.9	617.2	life advice
2	405	58	423.6	comedy	9	43	7.7	542.6	comedy
4	160	10.9	212.8	science	22	304	2.6	150.5	news
4	123	1.3	204.1	fashion	10	430	1.4	242.1	comedy
1	96	1.1	449	comedy	2	76	15.2	106.7	fashion
2	44	2.7	217.7	fashion	2	53	3.6	121.1	fashion
1	71	8	12.3	life advice	9	98	1	160.2	news
4	190	6.7	433.3	life advice	19	56	4.7	163.7	news
1	59	9.5	90.4	science	4	102	0.9	91.5	fashion
1	36	9.2	423.9	products	2	43	0.5	38.9	fashion
3	511	92.5	158.4	products	14	81	3.2	151.9	products
2	112	4.2	225.7	products	4	86	3.2	118.4	products
4	156	32.4	140.8	comedy	10	90	2.6	617.2	products

(a) Run the zero-truncated negative binomial model to regress the number of new videos on the other variables. Write the predicted model. What predictors turn out to be significant at the 5% level?

(b) Does the model have a good fit?

(c) Interpret estimated significant regression coefficients, if possible.

(d) Find the predicted number of new videos for a vlogger who posted a total of 87 videos on popular science, has 50,000 subscribers, and 254,000 views.

EXERCISE 6.6. A large insurance company that offers a range of insurance policies would like to model the number of insurance claims submitted within a five-year period. The data are retrieved for a random sample of 40 policyholders who have been with the company for at least ten years. The data include the number of claims made in the past five years, the number of claims made during the previous five years, current age, and gender of each policyholder. The data are given in the table below.

N claims past 5 years	N claims previous 5 years	Age	Gender	N claims past 5 years	N claims previous 5 years	Age	Gender
1	1	39	M	8	3	69	F
1	2	56	M	0	2	70	M
7	0	56	M	7	2	70	M
3	4	43	F	3	1	54	F
4	1	42	F	2	1	38	M
4	2	52	M	3	1	50	F
0	0	39	F	0	1	62	M
4	6	68	M	8	2	54	M
6	1	41	F	0	0	59	M
0	1	54	F	0	1	61	F
4	2	50	F	0	0	69	F
6	4	57	M	8	3	57	F
5	4	47	F	0	0	57	M
1	2	43	M	8	5	72	F
1	1	36	M	0	2	42	M
1	2	55	F	5	2	42	F
5	5	57	F	7	2	66	M
8	5	53	M	7	4	53	M
0	1	72	M	6	0	52	M
0	1	67	F	3	7	57	F

(a) Fit a zero-inflated negative binomial regression to model the number of claims made in the past five years. Model the probability of structural absence of claims as a function of the number of claims made in the previous five years. Model the positive responses as related to age and gender. What predictors are significant at the 5% level?

(b) Interpret the estimated significant coefficients.

(c) How good is the model fit? Give a quantitative answer.

(d) What is the predicted number of claims made in the past five years by a 55-year-old female policyholder who has made no claims in the previous five years?

EXERCISE 6.7. The aim of the study conducted by dental researchers is to model decayed, missing, and filled teeth (DMFT) index data for people in their 20s and 30s. They considered dental records for 38 randomly selected patients and picked age, gender, and oral hygiene as predictors. The data are presented in the table below.

DMFT index	Age	Gender	Oral hygiene	DMFT index	Age	Gender	Oral hygiene
1	28	F	low	1	29	M	low
2	30	F	low	0	30	M	med
0	26	M	high	0	25	F	high
2	25	M	low	1	22	F	low
2	30	F	med	7	37	M	med
2	19	M	med	2	30	F	med
0	24	F	high	0	23	M	high
2	27	F	med	0	21	M	high
0	18	F	high	5	25	F	low
3	21	F	med	0	25	F	high
10	39	M	low	2	28	M	low
1	28	M	high	4	25	M	low
1	24	F	med	0	19	M	med
2	26	F	med	3	23	F	low
1	23	F	high	1	20	F	med
2	24	F	low	0	30	M	high
1	21	M	med	6	33	M	low
2	30	M	med	3	23	M	med
0	31	F	med	2	35	M	med

(a) Fit a zero-inflated negative binomial model, regressing the probability of structural zeros of DMFT index on levels of oral hygiene. Regress positive observations of DMFT index on the demographic variables. Write down the predicted model. Discuss significance of predictors at the 5% significance level.
(b) Analyze the fit of the model.
(c) Give interpretation of the estimated significant coefficients.
(d) Find the predicted value of the DMFT index for a man, aged 28, with a high level of oral hygiene.

EXERCISE 6.8. Consider the data in Exercise 6.6.
(a) Fit a hurdle negative binomial regression to model the number of claims made in the past five years. Model the probability of zero claims as a function of the number of claims made in the previous five years. Model the positive responses as related to age and gender. Write down the fitted model explicitly. What predictors are significant at the 5% level?
(b) Discuss goodness-of-fit of the model.
(c) Interpret the estimated significant coefficients. What is the direction of the relationships?
(d) Find the predicted number of claims made in the past five years by a 55-year-old female policyholder who has made no claims in the previous five years.

EXERCISE 6.9. Researchers in sport medicine are interested in modeling the number of sport related injuries a collegiate athlete has during games throughout her/his career. They select 30 records that contain the total number of injuries (major or minor), athlete's gender, how many sports the athlete has participated in, and the number of injuries during practice. The researchers hypothesize that a lower number of injuries during practice might help explain zero injuries during games, while a larger number of sports might account for a larger number of injuries. They also hypothesize that female athletes have fewer injuries than males. The data are:

N game injuries	Gender	N sports	N practice injuries	N game injuries	Gender	N sports	N practice injuries
0	M	2	2	2	F	2	0
0	M	1	1	0	F	2	0
1	F	2	3	1	F	2	2
1	F	1	0	2	M	2	3
2	F	1	1	0	M	1	0
0	F	2	1	3	M	2	4
0	M	1	0	5	F	1	4
6	M	2	3	0	M	2	0
7	M	1	5	7	M	3	4
2	M	2	4	7	F	2	3
8	M	3	1	3	F	3	4
10	M	2	2	8	M	1	2
4	M	1	7	3	F	3	5
0	F	1	1	7	M	3	6
2	M	2	2	12	M	2	5

(a) Fit a hurdle negative binomial regression model. Specify the fitted model. Does it support the researchers' hypotheses? Discuss significance of predictor variables at the 5% significance level.

(b) Analyze the model fit.

(c) Give interpretation of the estimated significant coefficients.

(d) Calculate the predicted number of injuries for a male athlete who throughout his college years has participated in two sports, and who has received one minor injury during practice games.

Chapter 7

Regression Models for Proportion Response

Suppose the response variable y assumes continuous values in an interval $(0,1)$, where the endpoints may or may not be included. In this chapter we discuss four models that might be applied in this case: beta regression, zero-inflated beta regression (if zero is included in the range), one-inflated beta regression (if one is a possible observation), and zero-one-inflated beta regression (where both zero and one may be observed).

7.1 Beta Regression Model

7.1.1 Model Definition

Consider a response variable y that represents a proportion of events of interest (for instance, the proportion of books returned on-time in a library per week). We will suppose that y assumes values that range continuously between zero and one, not including the endpoints. The *beta regression model*[1] with predictors $x_1, \ldots, x_k$ prescribes that y follows a beta distribution with the probability density function

$$f(y) = \frac{y^{\mu\phi-1}(1-y)^{(1-\mu)\phi-1}}{B(\mu\phi,\,(1-\mu)\phi)}, \quad 0 < y < 1, \tag{7.1}$$

where the *location parameter*

$$\mu = \frac{\exp\{\beta_0 + \beta_1 x_1 + \cdots + \beta_k x_k\}}{1 + \exp\{\beta_0 + \beta_1 x_1 + \cdots + \beta_k x_k\}}, \tag{7.2}$$

---

[1]Originally proposed in Ferrari, S.L.P. and F. Cribari-Neto (2004). "Beta regression for modelling rates and proportions". *Journal of Applied Statistics*, 31(7): 799 – 815.

and the *dispersion* (or *scale*) *parameter* ϕ is a positive real number. It can be shown (see Exercise 7.1) that y has mean $\mathbb{E}(y) = \mu$ and variance $\mathbb{V}ar(y) = \dfrac{\mu(1 - \mu)}{1 + \phi}$.

7.1.2 Fitted Model

According to (7.2), the estimated parameters of the fitted model are

$$\widehat{\mu} = \frac{\exp\{\widehat{\beta}_0 + \widehat{\beta}_1\, x_1 + \cdots + \widehat{\beta}_k\, x_k\}}{1 + \exp\{\widehat{\beta}_0 + \widehat{\beta}_1\, x_1 + \cdots + \widehat{\beta}_k\, x_k\}} \quad \text{and} \quad \widehat{\phi}. \tag{7.3}$$

Technically speaking, the beta regression belongs to the class of generalized linear models. After a proper re-parametrization, it can be shown that the beta distribution belongs to the exponential family of distributions, and the mean response μ is related to the predictors through the logit link function.

7.1.3 Interpretation of Estimated Regression Coefficients

From (7.3), $\dfrac{\mu}{1 - \mu} = \exp\{\beta_0 + \beta_1\, x_1 + \cdots + \beta_k\, x_k\}$. Thus,

- for a unit increase in a numerical predictor x_1, the percent change in the estimated ratio $\dfrac{\widehat{\mu}}{1 - \widehat{\mu}} = \dfrac{\widehat{\mathbb{E}}(y)}{1 - \widehat{\mathbb{E}}(y)}$ is $(\exp\{\widehat{\beta}_1\} - 1) \cdot 100\%$, controlling for the other predictors.

- If x_1 is an indicator variable, then $\exp\{\widehat{\beta}_1\} \cdot 100\%$ represents the percent ratio of $\dfrac{\widehat{\mu}}{1 - \widehat{\mu}}$ for $x_1 = 1$ and that for $x_1 = 0$, keeping all the other predictors fixed.

7.1.4 Predicted Response

Since the mean of the response y is μ and using (7.3), we see that for some given predictors $x_1^0, \ldots, x_k^0$, the predicted response satisfies

$$y^0 = \frac{\exp\{\widehat{\beta}_0 + \widehat{\beta}_1\, x_1^0 + \cdots + \widehat{\beta}_k\, x_k^0\}}{1 + \exp\{\widehat{\beta}_0 + \widehat{\beta}_1\, x_1^0 + \cdots + \widehat{\beta}_k\, x_k^0\}}. \tag{7.4}$$

7.1.5 SAS Implementation

In SAS, a beta regression is estimated via the `glimmix` procedure (which stands for "generalized linear mixed" modeling). The syntax is:

```
proc glimmix data=data_name;
  class catpredictor1_name (ref='level_name') catpredictor2_name
        (ref='level_name') ...;
model response_name=<list of predictors>/dist=beta link=logit solution;
    output out=outdata pred(ilink)=predicted_name;
run;
```

- The option `solution` requests the estimates of the regression coefficients.
- SAS outputs the value of $\widehat{\phi}$, which is termed `Scale`.
- By default, SAS outputs the linear predictor $\widehat{\beta}_0 + \widehat{\beta}_1 x_1^0 + \cdots + \widehat{\beta}_k x_k^0$. The option (`ilink`) computes the predicted value of the response according to the formula (7.4).

7.1.6 R Implementation

The function `betareg()` in the library `betareg` fits the beta regression model. The syntax is

summary($fitted.model.name$<- betareg($response.name$~$x1.name$+$\cdots$ +$xk.name$, data=$data.name$, link="logit"))

- In the output, the estimate of the dispersion parameter is termed (`phi`).

7.1.7 Example

EXAMPLE 7.1. A professor of Library and Information Science has collected a random sample of 28 libraries and recorded total number of books each library has (in thousands), number of card holders (in thousands), library location (urban or rural), number of books checked out during a one-month period, and number of the books that were returned on-time, that is, on or prior to the due date. The professor has calculated the proportion of books returned on-time as the ratio between the number of books returned on-time and number of books checked out, and is interested in studying associations between this proportion and the three predictor variables. The code below fits the beta regression model.

```
data libraries;
input nbooks ncardholders location$ propontime @@;
cards;
9.8   0.7 rural 0.39 25.4 7.8 urban 0.81
14.7 1.7 urban 0.71 38.2 9.5 urban 0.85
35.5 5.8 urban 0.74 14.1 2.6 rural 0.47
16.3 2.9 rural 0.72 33.3 7.7 urban 0.83
19.9 3.1 rural 0.69 38.0 4.6 urban 0.76
44.1 8.3 urban 0.85 34.2 5.4 urban 0.86
12.7 2.7 rural 0.53 28.7 3.4 urban 0.78
9.5   1.0 rural 0.55 31.8 8.7 urban 0.78
21.1 6.2 urban 0.82 12.1 2.3 rural 0.42
26.3 0.9 rural 0.88 16.4 8.4 urban 0.88
11.8 1.6 rural 0.45 31.6 8.8 rural 0.64
24.6 4.5 rural 0.81 25.3 1.8 urban 0.73
12.4 1.9 rural 0.38 16.2 7.3 rural 0.66
13.7 6.2 urban 0.84 29.4 6.3 urban 0.72
;

proc glimmix;
 class location(ref='rural');
model propontime=nbooks ncardholders location/
   dist=beta link=logit solution;
run;
```

-2 Log Likelihood -53.40

Parameter Estimates

Effect	location	Estimate	Pr > |t|
Intercept		-0.2058	0.3896
nbooks		0.02624	0.0480
ncardholders		0.04494	0.2999
location	urban	0.5236	0.0316
location	rural	0	.
Scale		20.6478	

```
proc glimmix;
 model propontime=/dist=beta link=logit;
run;
```

-2 Log Likelihood -30.66

```
data deviance_test;
```

```
deviance = -30.66-(-53.40);
   pvalue=1-probchi(deviance,3);
run;

proc print;
run;
```

deviance   pvalue
 22.74 0.000045749

In the fitted model, the estimated mean has the form

$$\widehat{\mu} = \frac{\exp\{-0.2058 + 0.02624 * nbooks + 0.04494 * ncardholders + 0.5236 * urban\}}{1 + \exp\{-0.2058 + 0.02624 * nbooks + 0.04494 * ncardholders + 0.5236 * urban\}},$$

and the estimated dispersion parameter is $\widehat{\phi} = 20.6478$. The model fits the data well since the p-value of the corresponding test is tiny. The number of books and library location are significant predictors of the proportion of books returned on time at the 5% level. As the number of books increases by one thousand, the estimated ratio of the mean proportion of books returned on-time and the mean proportion of books not returned on-time, $\dfrac{\widehat{\mu}}{1-\widehat{\mu}}$, increases by $(\exp\{0.02624\} - 1) \cdot 100\% = 2.66\%$. This ratio for urban libraries is $\exp\{0.5236\} \cdot 100\% = 168.81\%$ of that for rural libraries.

To illustrate the calculations required for prediction, suppose we would like to predict proportion of books that are returned on time for a library in a rural area with 15,000 books and 2,500 card holders. We have

$$y^0 = \frac{\exp\{-0.2058 + 0.02624 * 15 + 0.04494 * 2.5\}}{1 + \exp\{-0.2058 + 0.02624 * 15 + 0.04494 * 2.5\}}$$

$$= \frac{\exp\{0.30015\}}{1 + \exp\{0.30015\}} = 0.57448.$$

The same prediction can be computed in SAS by typing:

```
data prediction;
input nbooks ncardholders location$;
cards;
15 2.5 rural
;

data libraries;
```

```
 set libraries prediction;
run;

proc glimmix;
 class location(ref='rural');
model propontime=nbooks ncardholders location /
   dist=beta link=logit solution;
 output out=outdata pred(ilink)=p_propontime;
run;

proc print data=outdata (firstobs=29 obs=29);
var p_propontime;
run;
```

p_propontime
     0.57447

The R code and output for this example are:

```
libraries.data<-read.csv(file="./Example7.1Data.csv",
header=TRUE, sep=",")
install.packages("betareg")
library(betareg)

#specifying reference category
location.rel<- relevel(libraries.data$location, ref="rural")

#fitting beta regression model
summary(fitted.model<- betareg(propontime ~ nbooks
+ ncardholders
+ location.rel, data=libraries.data, link="logit"))
```

	Estimate	Pr(>|z|)
(Intercept)	-0.20579	0.3729
nbooks	0.02624	0.0447
ncardholders	0.04494	0.2960
location.relurban	0.52357	0.0239
(phi)	20.648	

```
#checking model fit
intercept.only.model<- betareg(propontime ~ 1,
data=libraries.data, link="logit")
print(deviance<- -2*(logLik(intercept.only.model)
-logLik(fitted.model)))
```

22.73631

```
print(p.value<- pchisq(deviance, df=3, lower.tail=FALSE))
```

4.582989e-05

```
#using fitted model for prediction
print(predict(fitted.model, data.frame(nbooks=15,
ncardholders=2.5, location.rel="rural")))
```

0.5744721

□

7.2   Zero-inflated Beta Regression Model

7.2.1   Model Definition

When zero is a regular observation for the modeled proportion, a *zero-inflated
beta regression* may be used. In this model, the response variable y has a mixture of a beta distribution and a point mass at zero. That is, the distribution
function is modeled as

$$f(y) = \begin{cases} \pi_0, & \text{if } y = 0, \\ (1 - \pi_0) \dfrac{y^{\mu\phi-1}(1-y)^{(1-\mu)\phi-1}}{B(\mu\phi, (1-\mu)\phi)}, & \text{if } 0 < y < 1, \end{cases} \tag{7.5}$$

where the probability of zero

$$\pi_0 = \frac{\exp\{\beta_0 + \beta_1 x_1 + \cdots + \beta_m x_m\}}{1 + \exp\{\beta_0 + \beta_1 x_1 + \cdots + \beta_m x_m\}}, \tag{7.6}$$

the location parameter

$$\mu = \frac{\exp\{\gamma_0 + \gamma_1 x_{m+1} + \cdots + \gamma_{k-m} x_k\}}{1 + \exp\{\gamma_0 + \gamma_1 x_{m+1} + \cdots + \gamma_{k-m} x_k\}}, \tag{7.7}$$

and the dispersion parameter ϕ is a positive real number. Being a mixture of
beta distribution and a degenerate distribution at zero, this distribution doesn't
belong to the exponential family, and thus the model is not a generalized linear
model.

7.2.2 Fitted Model

By (7.6) and (7.7), the fitted model has the estimated parameters of the form:

$$\widehat{\pi}_0 = \frac{\exp\{\widehat{\beta}_0 + \widehat{\beta}_1\, x_1 + \cdots + \widehat{\beta}_m\, x_m\}}{1 + \exp\{\widehat{\beta}_0 + \widehat{\beta}_1\, x_1 + \cdots + \widehat{\beta}_m\, x_m\}}, \tag{7.8}$$

$$\widehat{\mu} = \frac{\exp\{\widehat{\gamma}_0 + \widehat{\gamma}_1\, x_{m+1} + \cdots + \widehat{\gamma}_{k-m}\, x_k\}}{1 + \exp\{\widehat{\gamma}_0 + \widehat{\gamma}_1\, x_{m+1} + \cdots + \widehat{\gamma}_{k-m}\, x_k\}}, \quad \text{and} \ \ \widehat{\phi}. \tag{7.9}$$

7.2.3 Interpretation of Estimated Regression Coefficients

If the two sets of predictor variables are different, the estimated beta and gamma coefficients yield the following interpretation. Estimated beta coefficients are interpreted as in the binary logistic regression in terms of odds in favor of observing a zero proportion (see Subsection 3.1.3). As for estimated gamma coefficients, for fixed values of $x_1, \ldots, x_k$, we can assume that π_0 is a constant, and, so μ can be viewed as the conditional mean response given that an observation is strictly above zero, that is, $\mu = \mathbb{E}(y|y > 0)$ (prove it!). Thus, assuming all the other x variables are unchanged,

- for a unit increase in a continuous x_{m+1}, the quantity $(\exp\{\widehat{\gamma}_1\} - 1) \cdot 100\%$

represents percent change in the estimated ratio $\dfrac{\widehat{\mu}}{1 - \widehat{\mu}} = \dfrac{\widehat{\mathbb{E}}(y|y > 0)}{1 - \widehat{\mathbb{E}}(y|y > 0)}$.

- If x_{m+1} is an indicator variable, then $\exp\{\widehat{\gamma}_1\} \cdot 100\%$ is interpreted as the difference in the estimated ratio for $x_1 = 1$ and that for $x_1 = 0$.

7.2.4 Predicted Response

Since $\mathbb{E}(y) = (1 - \pi_0)\,\mu$ (show it!), and using (7.8) and (7.9), we see that for a fixed set $x_1^0, \ldots, x_k^0$, the predicted response y^0 is found according to the formula:

$$y^0 = \left(1 - \frac{\exp\{\widehat{\beta}_0 + \widehat{\beta}_1\, x_1^0 + \cdots + \widehat{\beta}_m\, x_m^0\}}{1 + \exp\{\widehat{\beta}_0 + \widehat{\beta}_1\, x_1^0 + \cdots + \widehat{\beta}_m\, x_m^0\}}\right) \frac{\exp\{\widehat{\gamma}_0 + \widehat{\gamma}_1\, x_{m+1}^0 + \cdots + \widehat{\gamma}_{k-m}\, x_k^0\}}{1 + \exp\{\widehat{\gamma}_0 + \widehat{\gamma}_1\, x_{m+1}^0 + \cdots + \widehat{\gamma}_{k-m}\, x_k^0\}}$$

$$= \left(1 + \exp\{\widehat{\beta}_0 + \widehat{\beta}_1\, x_1^0 + \cdots + \widehat{\beta}_m\, x_m^0\}\right)^{-1} \frac{\exp\{\widehat{\gamma}_0 + \widehat{\gamma}_1\, x_{m+1}^0 + \cdots + \widehat{\gamma}_{k-m}\, x_k^0\}}{1 + \exp\{\widehat{\gamma}_0 + \widehat{\gamma}_1\, x_{m+1}^0 + \cdots + \widehat{\gamma}_{k-m}\, x_k^0\}}. \tag{7.10}$$

7.2.5 SAS Implementation

There is currently no ready procedure that would fit a zero-inflated beta regression in SAS. One would need to use the `nlmixed` procedure ("non-linear mixed") that allows to specify any log-likelihood function of the response variable. In this procedure, all predictors must be either numeric or 0-1 variables. We express the log-likelihood function in terms of the natural logarithm of the gamma function `lgamma`, that is, we write the log-likelihood function as: for $y = 0$, log-likelihood= $\ln \pi_0$, and, for $0 < y < 1$, log-likelihood=

$$\ln \left[(1-\pi_0) \frac{y^{\mu\phi-1}(1-y)^{(1-\mu)\phi-1}}{B(\mu\phi, (1-\mu)\phi)} \right] = \ln \left[(1-\pi_0) \frac{\Gamma(\phi)}{\Gamma(\mu\phi)\Gamma((1-\mu)\phi)} y^{\mu\phi-1}(1-y)^{(1-\mu)\phi-1} \right] = \ln(1-\pi_0) + \ln\Gamma(\phi) - \ln\Gamma(\mu\phi) - \ln\Gamma((1-\mu)\phi) + (\mu\phi-1)\ln y + ((1-\mu)\phi-1)\ln(1-y).$$

The syntax for fitting the zero-inflated beta regression and predicting response is below. It is assumed that the last row in the data set contains the values for prediction.

```
proc nlmixed data=data_name;
    parms b0=init_value ... bm=init_value g0=init_value ...
          g{k-m}=init_value phi=init_value;
      pi0 = exp(b0+b1*x1_name+...+bm*xm_name)/
            (1 + exp(b0+b1*x1_name +...+bm*xm_name));
    mu = exp(g0+g1*x{m+1}_name+...+g{k-m}*xk_name)/
          (1 + exp(g0+g1*x{m+1}_name+...+g{k-m}*xk_name));
if (response_name=0) then loglikelihood_name=log(pi0);
  else loglikelihood_name =log(1-pi0)+lgamma(phi)-lgamma(mu*phi)
    -lgamma((1-mu)*phi)+(mu*phi-1)*log(response_name)
      +((1-mu)*phi-1)*log(1-response_name);
model response_name ~ general(loglikelihood_name);
    predict (1-pi0)*mu out=outdata_name;
run;
```

• In the `parms` statement, the initial values for the variables b0,..., bm, g0,..., g{k-m}, and phi can be chosen arbitrarily.
• The predicted response can be found in the last row of the data set *outdata_name*.

7.2.6 R Implementation

In R, the function `gamlss()` in the library `gamlss` can be applied to fit a zero-inflated beta regression. Here "gamlss" stands for "generalized additive models for location, scale, and shape". R uses the term `mu` for the location parameter μ, `nu` for the probability of zero π_0, and `sigma` for the log of the dispersion parameter ϕ, that is, `sigma`$= \ln \phi$. The syntax is

`summary`(*fitted.model.name*`<- gamlss`(*response.name* $\sim$ *x{m+1}.name* + $\cdots$ + *xk.name*, `mu.link="logit"`, `nu.formula=`$\sim$ *x1.name* + $\cdots$ + *xm.name*, `nu.link="logit"`, `data=`*data.name*, `family=BEZI`))

If one wishes to predict response for a new data point, the quickest way to do that is to use function `predictAll()` as shown below:

parampred.name`<- predictAll`(*fitted.model*, `newdata`
`=data.frame`(*catpredictor1.name*`="`*value*`"`, $\ldots$,
numpredictork.name=value), `type="response"`)

The *parampred.name* contains predicted values for parameters `mu` and `nu`, thus the predicted response can be retrieved by specifying the expression

`print`(*pred.name*`<- (1-`*parampred.name*`$nu`)`*`*parampred.name*`$mu`)

- It is important to know that the variables in `newdata` must match the variables in the original data *data_name*. It means that prior to running the function `gamlss()`, all the predictor variables that are computed or modified, must be added to the data set *data.name*. This can be accomplished by giving them names that reference the data set, *data.name$variable.name*.

7.2.7 Example

EXAMPLE 7.2. A college Parking and Transportation Services conducted a survey regarding the mode of transportation to campus in the past four weeks. For this analysis, a stratified sample of 60 respondents was drawn for the purpose of oversampling people who sometimes bike to campus. The variables recorded are the number of trips to campus in the past four weeks, the number of times the respondent biked to campus, respondent's status (student/faculty/staff), respondent's gender (F/M), duration of parking permit (6, 9, or 12-month, modeled as a continuous variable), and distance to campus. The code below fits a zero-inflated beta regression to the proportion of time a respondent biked to

176

campus. Gender and distance to campus were used as predictors of π_0, whereas status and parking permit were used to model μ.

```
data transport;
input ntrips nbiked status$ gender$ parking distance @@;
  propbiked=nbiked/ntrips;
   faculty=(status='faculty');
    staff=(status='staff');
 male=(gender='M');
cards;
26 6   student F 6   7    13 0  faculty M 9   15   17 0  faculty M 9   31
15 0   faculty M 9   9    17 0  student M 6   34   20 13 staff   F 6   3
17 0   student F 9   17   14 0  faculty F 9   10   26 7  student M 6   3
8  5   faculty F 12 3    17 0  student F 12 22   18 0  student M 12 8
15 0   staff   F 12 4    17 7  faculty F 12 5    20 15 staff   M 9   6
8  0   student F 6   5    26 9  student M 9   3    12 11 faculty F 6   12
13 5   student M 9   7    8  6  faculty M 9   5    9  0  student M 9   7
12 2   faculty F 9   8    11 0  student F 6   13   8  1  staff   F 9   16
14 0   faculty M 12 35   9  2  staff   F 12 8    19 12 student M 6   5
20 0   faculty M 12 60   28 11 staff   M 9   1    10 0  faculty F 6   30
14 4   staff M 12   4    8  5  student F 6   2    9  0  staff   F 12 16
12 0   faculty M 12 15   12 0  student F 9   25   21 19 faculty M 9   3
23 22 faculty F 6   14   27 22 faculty M 9   6    14 3  student F 6   12
15 9   faculty M 12 3    8  2  student F 9   6    15 5  faculty F 12 6
12 10 student F 6   3    8  5  faculty M 9   4    14 4  student F 6   2
15 0   student M 6   10   12 7  student F 6   6    23 7  staff   M 12 2
18 13 student F 6   2    9  0  student M 6   16   10 9  faculty M 9   6
8  1   staff   M 12 7    12 2  student F 6   12   14 0  staff   M 12 7
8  7   student F 12 12   16 14 student M 6   2    11 4  faculty F 12 5
10 0   faculty F 12 22   16 0  staff   M 12 4    10 0  faculty M 12 11
;

proc nlmixed;
 parms b0=0.1 b1=0.1 b2=0.1 g0=0.1 g1=0.1 g2=0.1 g3=0.1 phi=0.1;
  pi0 = exp(b0+b1*male+b2*distance)/(1+exp(b0+b1*male+b2*distance));
     mu = exp(g0+g1*faculty+g2*staff+g3*parking)/(1+
             exp(g0+g1*faculty+g2*staff+g3*parking));
   if (propbiked=0) then loglikelihood=log(pi0);
       else loglikelihood =log(1-pi0)+lgamma(phi)-lgamma(mu*phi)
             -lgamma((1-mu)*phi)+(mu*phi-1)*log(propbiked)
                  +((1-mu)*phi-1)*log(1-propbiked);
```

```
        model propbiked~general(loglikelihood);
run;
```

-2 Log Likelihood 33.3

Parameter Estimates

Parameter	Estimate	Pr > |t|
b0	-4.4897	0.0006
b1	1.7728	0.0484
b2	0.3396	0.0009
g0	1.1040	0.0665
g1	1.0635	0.0127
g2	0.03280	0.9455
g3	-0.1638	0.0475
phi	3.9871	

```
proc nlmixed;
  parms b0=0.1 g0=0.1 phi=0.1;
    pi0 = exp(b0)/(1+ exp(b0));
        mu = exp(g0)/(1+exp(g0));
    if (propbiked=0) then loglikelihood=log(pi0);
        else loglikelihood =log(1-pi0)+lgamma(phi)-lgamma(mu*phi)
            -lgamma((1-mu)*phi)+(mu*phi-1)*log(propbiked)
            +((1-mu)*phi-1)*log(1-propbiked);
    model propbiked~general(loglikelihood);
run;
```

-2 Log Likelihood 77.5

```
data deviance_test;
  deviance=77.5-33.3;
    pvalue=1-probchi(deviance,5);
run;

proc print;
run;
```

deviance	pvalue
44.2	2.1095E-8

In the fitted model, the estimates of the parameters are

$$\widehat{\pi}_0 = \frac{\exp\{-4.4897 + 1.7728 * male + 0.3396 * distance\}}{1 + \exp\{-4.4897 + 1.7728 * male + 0.3396 * distance\}},$$

178

$$\widehat{\mu} = \frac{\exp\{1.1040 + 1.0635 * faculty + 0.03280 * staff - 0.1638 * parking\}}{1 + \exp\{1.1040 + 1.0635 * faculty + 0.03280 * staff - 0.1638 * parking\}},$$

and $\widehat{\phi} = 3.9871$. The model fits the data well as evidenced by a tiny p-value in the deviance test. At the 5% significance level, male and distance are significant predictors of π_0, whereas faculty and parking significantly predict μ. The estimated odds in favor of not biking to campus for males are $\exp\{1.7728\} \cdot 100\% = 588.73\%$ of those for females, and for a one-mile increase in distance, these estimated odds increase by $(\exp\{0.3396\}-1)\cdot100\% = 40.44\%$. The estimated ratio $\dfrac{\widehat{\mu}}{1-\widehat{\mu}}$ for faculty is $\exp\{1.0635\} \cdot 100\% = 289.65\%$ of that for students. For a one-month increase in duration of parking permit, the estimated ratio changes by $(\exp\{-0.1638\} - 1) \cdot 100\% = -15.11\%$, that is, it decreases by 15.11%.

To predict the response for a female student who lives 3 miles away from campus and who purchases parking permit for 6 months during a year. Utilizing (7.10), we have

$$y^0 = \left(1+\exp\{-4.4897+0.3396*3\}\right)^{-1} \frac{\exp\{1.1040 - 0.1638 * 6\}}{1 + \exp\{1.1040 - 0.1638 * 6\}} = 0.5143.$$

In SAS, the following lines of code produce the same predicted value. We present only the relevant quantities in the output.

```
data prediction;
input parking distance faculty staff male;
cards;
6 3 0 0 0
;

data transport;
 set transport prediction;
run;

proc nlmixed;
 parms b0=0.1 b1=0.1 b2=0.1 g0=0.1 g1=0.1 g2=0.1
       g3=0.1 phi=0.1;
  pi0 = exp(b0+b1*male+b2*distance)/(1+
    exp(b0+b1*male+b2*distance));
   mu = exp(g0+g1*faculty+g2*staff+g3*parking)/(1+
     exp(g0+g1*faculty+g2*staff+g3*parking));
   if (propbiked=0) then loglikelihood=log(pi0);
   else loglikelihood =log(1-pi0)+lgamma(phi)-lgamma(mu*phi)
```

```
        -lgamma((1-mu)*phi)+(mu*phi-1)*log(propbiked)
            +((1-mu)*phi-1)*log(1-propbiked);
    model propbiked~general(loglikelihood);
        predict (1-pi0)*mu out=outdata;
run;

proc print data=outdata (firstobs=61 obs=61);
var Pred;
run;
```

```
    Pred
0.51433
```

Finally, the script and relevant output in R are given below.

```
transport.data<-read.csv(file="./Example7.2Data.csv",
header=TRUE, sep=",")
install.packages("gamlss")
library(gamlss)

#computing new variables and making them part
#of the original data set
transport.data$propbiked<- transport.data
$nbiked/transport.data$ntrips
transport.data$faculty
<-ifelse(transport.data$status=="faculty",1,0)
transport.data$staff
<-ifelse(transport.data$status=="staff",1,0)
transport.data$male
<-ifelse(transport.data$gender=="M",1,0)

#fitting zero-inflated beta regression model
summary(fitted.model<- gamlss
(propbiked  ~ faculty + staff + parking,
mu.link="logit", nu.formula= ~ male + distance, nu.link="logit",
data=transport.data, family=BEZI))
```

Mu link function: logit
Mu Coefficients:

	Estimate	Pr(>|t|)
(Intercept)	1.10390	0.0673
faculty	1.06338	0.0131
staff	0.03265	0.9458
parking	-0.16374	0.0482

Sigma link function: log
Sigma Coefficients:
          Estimate
(Intercept) 1.3831

Nu link function: logit
Nu Coefficients:
            Estimate   Pr(>|t|)
(Intercept) -4.48970   0.000699
male   1.77277   0.049067
distance   0.33957   0.000960

```
#checking model fit
intercept.only.model<- gamlss(propbiked  ~ 1, mu.link="logit",
nu.formula= ~ 1, nu.link="logit", data=transport.data,
family=BEZI)
print(deviance<- -2*(logLik(intercept.only.model)
-logLik(fitted.model)))
```

44.18192

```
print(p.value<- pchisq(deviance, df=5, lower.tail=FALSE))
```

2.127389e-08

```
#using fitted model for prediction
param.pred<- predictAll(fitted.model, newdata
=data.frame(parking=6, distance=3, faculty=0, staff=0,
male=0), type="response")
print((1-param.pred$nu)*param.pred$mu)
```

0.5143409

□

7.3   One-inflated Beta Regression Model

7.3.1   Model Definition

Suppose y variable assumes values in the interval $(0,1]$, that is, 1 is a possible observation. Then the data may be modeled by a *one-inflated beta regression*. In this model, the response variable y has a mixture beta distribution with a

181

point mass at one. That is, the distribution function of y is

$$
f(y) = \begin{cases} (1 - \pi_1) \dfrac{y^{\mu\phi-1} (1 - y)^{(1-\mu)\phi-1}}{B(\mu\phi,\, (1-\mu)\phi)}, & \text{if } 0 < y < 1, \\[2ex] \pi_1, & \text{if } y = 1. \end{cases}
$$

Here the probability of one

$$
\pi_1 = \frac{\exp\{\beta_0 + \beta_1\, x_1 + \cdots + \beta_m\, x_m\}}{1 + \exp\{\beta_0 + \beta_1\, x_1 + \cdots + \beta_m\, x_m\}}, \tag{7.11}
$$

the location parameter

$$
\mu = \frac{\exp\{\gamma_0 + \gamma_1\, x_{m+1} + \cdots + \gamma_{k-m}\, x_k\}}{1 + \exp\{\gamma_0 + \gamma_1\, x_{m+1} + \cdots + \gamma_{k-m}\, x_k\}}, \tag{7.12}
$$

and the dispersion parameter ϕ is a positive real number.

Analogously to zero-inflated beta regression, since the underlying distribution is a mixture distribution, this model is not a generalized linear regression model.

7.3.2 Fitted Model

In view of expressions (7.11) and (7.12), the fitted model has the estimated parameters

$$
\widehat{\pi}_1 = \frac{\exp\{\widehat{\beta}_0 + \widehat{\beta}_1\, x_1 + \cdots + \widehat{\beta}_m\, x_m\}}{1 + \exp\{\widehat{\beta}_0 + \widehat{\beta}_1\, x_1 + \cdots + \widehat{\beta}_m\, x_m\}}, \tag{7.13}
$$

$$
\widehat{\mu} = \frac{\exp\{\widehat{\gamma}_0 + \widehat{\gamma}_1\, x_{m+1} + \cdots + \widehat{\gamma}_{k-m}\, x_k\}}{1 + \exp\{\widehat{\gamma}_0 + \widehat{\gamma}_1\, x_{m+1} + \cdots + \widehat{\gamma}_{k-m}\, x_k\}}, \quad \text{and } \widehat{\phi}. \tag{7.14}
$$

7.3.3 Interpretation of Estimated Regression Coefficients

The estimated beta coefficients are interpreted as in the logistic regression in terms of odds in favor of observing a one (see Subsection 3.1.3). As concerns the estimated gamma coefficients, for fixed values of $x_1, \ldots, x_m$, we can assume π_1 is a constant, and so, μ represents the conditional mean response given that an observation is below one, that is, $\mu = \mathbb{E}(y|y < 1)$ (show it!). Therefore, under the assumption that all the other predictors stay fixed,

• when a continuous x_{m+1} increases by one unit, the percent change in the estimated ratio $\dfrac{\widehat{\mu}}{1 - \widehat{\mu}} = \dfrac{\widehat{\mathbb{E}}(y|y < 1)}{1 - \widehat{\mathbb{E}}(y|y < 1)}$ is $(\exp\{\widehat{\gamma}_1\} - 1) \cdot 100\%$.

• For an indicator variable x_{m+1}, the quantity $\exp\{\widehat{\gamma}_1\} \cdot 100\%$ represents the percent ratio of $\dfrac{\widehat{\mu}}{1 - \widehat{\mu}}$ for $x_1 = 1$ and that for $x_1 = 0$.

7.3.4 Predicted Response

It can be proven (do it!) that $\mathbb{E}(y) = \pi_1 + (1 - \pi_1)\,\mu$. Thus, utilizing (7.13) and (7.14), we find that for given $x_1^0, \ldots, x_k^0$, the predicted response y^0 satisfies:

$$y^0 = \frac{\exp\{\widehat{\beta}_0 + \widehat{\beta}_1\,x_1^0 + \cdots + \widehat{\beta}_m\,x_m^0\}}{1 + \exp\{\widehat{\beta}_0 + \widehat{\beta}_1\,x_1^0 + \cdots + \widehat{\beta}_m\,x_m^0\}} + \left(1 - \frac{\exp\{\widehat{\beta}_0 + \widehat{\beta}_1\,x_1^0 + \cdots + \widehat{\beta}_m\,x_m^0\}}{1 + \exp\{\widehat{\beta}_0 + \widehat{\beta}_1\,x_1^0 + \cdots + \widehat{\beta}_m\,x_m^0\}}\right)$$

$$\times \frac{\exp\{\widehat{\gamma}_0 + \widehat{\gamma}_1\,x_{m+1}^0 + \cdots + \widehat{\gamma}_{k-m}\,x_k^0\}}{1 + \exp\{\widehat{\gamma}_0 + \widehat{\gamma}_1\,x_{m+1}^0 + \cdots + \widehat{\gamma}_{k-m}\,x_k^0\}} = \left(1 + \exp\{\widehat{\beta}_0 + \widehat{\beta}_1\,x_1^0 + \cdots + \widehat{\beta}_m\,x_m^0\}\right)^{-1} \times$$

$$\times \left(\exp\{\widehat{\beta}_0 + \widehat{\beta}_1\,x_1^0 + \cdots + \widehat{\beta}_m\,x_m^0\} + \frac{\exp\{\widehat{\gamma}_0 + \widehat{\gamma}_1\,x_{m+1}^0 + \cdots + \widehat{\gamma}_{k-m}\,x_k^0\}}{1 + \exp\{\widehat{\gamma}_0 + \widehat{\gamma}_1\,x_{m+1}^0 + \cdots + \widehat{\gamma}_{k-m}\,x_k^0\}}\right). \quad (7.15)$$

7.3.5 SAS Implementation

The procedure `nlmixed` is used to fit a one-inflated beta regression. The syntax is:

```
proc nlmixed data=data_name;
   parms b0=init_value ... bm=init_value g0=init_value ...
        g{k-m}=init_value phi=init_value;
      pi1 = exp(b0+b1*x1_name+...+bm*xm_name)/
            (1 + exp(b0+b1*x1_name +...+bm*xm_name));
     mu = exp(g0+g1*x{m+1}_name+...+g{k-m}*xk_name)/
           (1 + exp(g0+g1*x{m+1}_name+...+g{k-m}*xk_name));
 if (response_name=1) then loglikelihood_name=log(pi1);
   else loglikelihood_name =log(1-pi1)+lgamma(phi)-lgamma(mu*phi)
    -lgamma((1-mu)*phi)+(mu*phi-1)*log(response_name)
     +((1-mu)*phi-1)*log(1-response_name);
 model response_name ~ general(loglikelihood_name);
    predict pi1+(1-pi1)*mu out=outdata_name;
 run;
```

7.3.6 R Implementation

In R, the function `gamlss()` in the library `gamlss` fits a one-inflated beta regression. In the output, `mu` is the location parameter μ, `nu` is the probability of one π_1, and `sigma`$= \ln \phi$. The syntax is

```
summary(fitted.model.name<- gamlss(response.name ~ x{m+1}.name + ···
+ xk.name, mu.link="logit", nu.formula=~x1.name+···+xm.name,
```

```
nu.link="logit", data =data.name, family=BEOI))
```

parampred.name<- predictAll(*fitted.model*, newdata
=data.frame(*catpredictor1.name*="*value*", ...,
numpredictork.name=*value*), type="response")

print(*parampred.name*$nu+(1-*parampred.name*$nu)*parampred.name$mu)

- Variables listed in `newdata` must be contained in the original data set *data_name*.

7.3.7   Example

EXAMPLE 7.3. A medical student is studying medication adherence in patients with high cholesterol. He collects data on patients' age (in years), gender (M/F), depression (1=yes/0=no), diabetes (1=yes/0=no), and total number of medications taken daily. In addition, he obtains pharmaceutical records and calculates proportion of days covered (pdc) for cholesterol-lowering medication, that is, the proportion of days that the patient has taken the medication. For patients who consistently adhere to their medication regimen, proportion of days covered is equal to one, and therefore fitting one-inflated regression would be appropriate. Below are the statements and outputs in SAS. In what follows π_1 is regressed on diabetes and number of medications, whereas μ is modeled as a function of age, gender, and depression.

```
data medadherence;
input age gender$ depression diabetes nmeds pdc@@;
female=(gender='F');
cards;
62 M 0 1 2 1.00   73 M 1 0 5 0.96   56 M 0 0 5 0.08
58 M 0 0 4 0.32   56 F 0 0 3 1.00   62 F 0 0 4 0.86
31 F 0 1 5 0.62   59 M 0 0 5 0.23   64 M 0 1 3 0.56
67 F 0 0 4 0.91   59 F 0 0 3 0.39   90 M 1 0 4 0.86
83 M 0 1 2 1.00   70 F 0 0 2 0.41   58 M 1 0 6 0.36
61 F 0 0 2 0.85   70 F 1 1 4 0.87   56 M 0 0 3 0.28
67 F 0 0 3 0.80   60 F 0 0 3 0.16   67 M 0 1 3 1.00
67 M 1 0 3 0.24   71 M 1 0 2 0.64   67 F 1 1 2 1.00
46 M 0 0 3 0.48   62 F 0 0 6 0.32   63 M 0 1 1 0.27
67 F 0 1 4 0.92   83 M 0 0 2 0.83   62 M 0 1 4 0.47
71 F 0 1 3 1.00   49 F 1 0 5 0.26   66 M 0 0 1 1.00
```

```
61 M 0 0 4 0.51   66 F 0 0 4 0.66   71 M 0 0 5 0.82
58 F 0 0 3 0.40   70 F 0 1 4 0.88   83 M 0 1 2 1.00
64 M 0 0 5 0.39
;

proc nlmixed;
 parms b0=0.1 b1=0.1 b2=0.1 g0=0.1 g1=0.1 g2=0.1 g3=0.1 phi=0.1;
  pi1 = exp(b0+b1*diabetes+b2*nmeds)/
  (1+exp(b0+b1*diabetes+b2*nmeds));
    mu = exp(g0+g1*age+g2*depression+g3*female)/(1+
       exp(g0+g1*age+g2*depression+g3*female));
      if (pdc=1) then loglikelihood=log(pi1);
        else loglikelihood =log(1-pi1)+lgamma(phi)-lgamma(mu*phi)
         -lgamma((1-mu)*phi)+ (mu*phi-1)*log(pdc)
         +((1-mu)*phi-1)*log(1-pdc);
    model pdc ~ general(loglikelihood);
run;
```

-2 Log Likelihood 6.4

Parameter Estimates

Parameter	Estimate	Pr > |t|
b0	1.5679	0.3348
b1	2.4896	0.0327
b2	-1.4320	0.0214
g0	-3.9846	0.0041
g1	0.06073	0.0045
g2	0.1210	0.7489
g3	0.7162	0.0281
phi	4.5022	

```
proc nlmixed;
 parms b0=0.1 g0=0.1 phi=0.1;
     pi1 = exp(b0)/(1+exp(b0));
    mu = exp(g0)/(1+exp(g0));
     if (pdc=1) then loglikelihood=log(pi1);
    else loglikelihood =log(1-pi1)+lgamma(phi)-lgamma(mu*phi)
 -lgamma((1-mu)*phi)+(mu*phi-1)*log(pdc)
    +((1-mu)*phi-1)*log(1-pdc);
    model pdc~general(loglikelihood);
run;
```

-2 Log Likelihood 36.4

```
data deviance_test;
 deviance=36.4-6.4;
   pvalue=1-probchi(deviance,5);
run;

proc print;
run;
```

```
deviance      pvalue
   30   0.000014749
```

The parameters of the fitted model are

$$\widehat{\pi}_1 = \frac{\exp\{1.5679 + 2.4896 * diabetes - 1.4320 * nmeds\}}{1 + \exp\{1.5679 + 2.4896 * diabetes - 1.4320 * nmeds\}},$$

$$\widehat{\mu} = \frac{\exp\{-3.9846 + 0.06073 * age + 0.7162 * depression + 0.1210 * female\}}{1 + \exp\{-3.9846 + 0.06073 * age + 0.7162 * depression + 0.1210 * female\}},$$

and $\widehat{\phi} = 4.5022$. This model has a good fit due to a low p-value in the goodness-of-fit test. Diabetes and number of medications are significant predictors of probability of one, whereas age and female are significant predictors of μ, at the 0.05 level of significance. For patients with diabetes, the estimated odds in favor of perfect medication adherence are $(\exp\{2.4896\}) \cdot 100\% = 1,205.65\%$ of those without diabetes. If the number of medications increases by one, the estimated odds in favor of 100%-adherence change by $(\exp\{-1.4320\} - 1) \cdot 100\% = -76.12\%$, that is, decreases by 76.12%. When age increases by one year, the estimated ratio $\dfrac{\widehat{\mu}}{1 - \widehat{\mu}} = \dfrac{\widehat{\mathbb{E}}(y|y < 1)}{1 - \widehat{\mathbb{E}}(y|y < 1)}$ increases by $(\exp\{0.06073\} - 1) \cdot 100\% = 6.26\%$. For females, this estimated ratio is $\exp\{0.1210\} \cdot 100\% = 112.86\%$ of that for males.

Further, according to (7.15), the predicted proportion of days covered for a 77-year old female patient who has no depression but has diabetes and who takes a total of three medications is computed as

$$y^0 = \left(1 + \exp\{1.5679 + 2.4896 - 1.4320 * 3\}\right)^{-1} \times$$

$$\times \left(\exp\{1.5679 + 2.4896 - 1.4320 * 3\} + \frac{\exp\{-3.9846 + 0.06073 * 77 + 0.7162\}}{1 + \exp\{-3.9846 + 0.06073 * 77 + 0.7162\}}\right)$$

$$= 0.8900.$$

SAS calculates the same predicted value by running the following statements.

```
data prediction;
input age depression diabetes nmeds female;
cards;
77 0 1 3 1
;

data medadherence;
 set medadherence prediction;
run;

proc nlmixed;
 parms b0=0.1 b1=0.1 b2=0.1 g0=0.1 g1=0.1
      g2=0.1 g3=0.1 phi=0.1;
   pi1 = exp(b0+b1*diabetes+b2*nmeds)/(1+
     exp(b0+b1*diabetes+b2*nmeds));
   mu = exp(g0+g1*age+g2*depression+g3*female)/(1+
      exp(g0+g1*age+g2*depression+g3*female));
        if (pdc=1) then loglikelihood=log(pi1);
   else loglikelihood =log(1-pi1)+lgamma(phi)-lgamma(mu*phi)
      -lgamma((1-mu)*phi)+(mu*phi-1)*log(pdc)
      +((1-mu)*phi-1)*log(1-pdc);
     model pdc ~ general(loglikelihood);
         predict pi1+(1-pi1)*mu out=outdata;
run;

proc print data=outdata (firstobs=41 obs=41);
var Pred;
run;

    Pred
0.89003
```

The R script and output for this example are presented below.

```
medadherence.data<-read.csv(file="./Example7.3Data.csv",
header=TRUE, sep=",")
install.packages("gamlss")
library(gamlss)

#computing new variable and making it part of the original data set
medadherence.data$female<- ifelse(medadherence.data$gender=="F",1,0)
```

```
#fitting one-inflated beta regression model
summary(fitted.model<- gamlss(pdc ~ age + female + depression,
mu.link="logit", nu.formula= ~ diabetes + nmeds,
  nu.link="logit",
data=medadherence.data, family=BEOI))
```

Mu link function: logit
Mu Coefficients:

	Estimate	Pr(>|t|)
(Intercept)	-3.97514	0.00469
age	0.06058	0.00511
female	0.71620	0.02951
depression	0.12123	0.74883

Sigma link function: log
Sigma Coefficients:

	Estimate
(Intercept)	1.504

Nu link function: logit
Nu Coefficients:

	Estimate	Pr(>|t|)
(Intercept)	1.5679	0.3362
diabetes	2.4896	0.0342
nmeds	-1.4320	0.0227

```
#checking model fit
intercept.only.model<- gamlss(pdc ~ 1, mu.link="logit",
nu.formula= ~ 1, nu.link="logit", data=medadherence.data,
family=BEOI)
print(deviance<- -2*(logLik(intercept.only.model)
-logLik(fitted.model)))
```

29.97312

```
print(p.value<- pchisq(deviance, df=5, lower.tail=FALSE))
```

1.492935e-05

```
#using fitted model for prediction
param.pred<- predictAll(fitted.model, newdata=data.frame(age=77,
female=1, depression=0, diabetes=1, nmeds=3), type="response")
print(param.pred$nu+(1-param.pred$nu)*param.pred$mu)
```

0.889879

□

7.4 Zero-one-inflated Beta Regression Model

7.4.1 Model Definition

In the *zero-one-inflated beta regression model*, the response variable y changes according to a distribution with the density:

$$f(y) = \begin{cases} \pi_0, & \text{if } y = 0, \\ (1 - \pi_0 - \pi_1) \frac{y^{\mu\phi-1}(1-y)^{(1-\mu)\phi-1}}{B(\mu\phi, (1-\mu)\phi)}, & \text{if } 0 < y < 1, \\ \pi_1, & \text{if } y = 1. \end{cases}$$

Here the parameters depend on predictors $x_1, \ldots, x_k$ via the following relations:

$$\mu = \frac{\exp\{\beta_0 + \beta_1 x_1 + \cdots + \beta_m x_m\}}{1 + \exp\{\beta_0 + \beta_1 x_1 + \cdots + \beta_m x_m\}}, \tag{7.16}$$

$$\pi_0 = \frac{\nu}{1 + \nu + \tau}, \quad \text{and} \quad \pi_1 = \frac{\tau}{1 + \nu + \tau}, \tag{7.17}$$

where

$$\nu = \exp\{\gamma_0 + \gamma_1 x_{m+1} + \cdots + \gamma_{l-m} x_l\} \quad \text{and} \quad \tau = \exp\{\zeta_0 + \zeta_1 x_{l+1} + \cdots + \zeta_{k-l} x_k\}, \tag{7.18}$$

and the dispersion parameter ϕ is a positive constant. Note that the probabilities of zero and one cannot be modeled completely independently of each other, and the given parametrization makes the regression parameters estimable.

Also, similarly to zero- and one-inflated beta regressions, in this situation the distribution of the response is a mixture distribution, and thus, the regression doesn't belong to the class of generalized linear models.

7.4.2 Fitted Model

Taking into account (7.16)-(7.18), the estimated parameters in the fitted model can be written as:

$$\widehat{\mu} = \frac{\exp\{\widehat{\beta}_0 + \widehat{\beta}_1 x_1 + \cdots + \widehat{\beta}_m x_m\}}{1 + \exp\{\widehat{\beta}_0 + \widehat{\beta}_1 x_1 + \cdots + \widehat{\beta}_m x_m\}}, \tag{7.19}$$

$$\widehat{\pi}_0 = \frac{\widehat{\nu}}{1 + \widehat{\nu} + \widehat{\tau}}, \quad \text{and} \quad \widehat{\pi}_1 = \frac{\widehat{\tau}}{1 + \widehat{\nu} + \widehat{\tau}}, \tag{7.20}$$

where

$$\widehat{\nu} = \exp\{\widehat{\gamma}_0 + \widehat{\gamma}_1 x_{m+1} + \cdots + \widehat{\gamma}_{l-m} x_l\}, \quad \widehat{\tau} = \exp\{\widehat{\zeta}_0 + \widehat{\zeta}_1 x_{l+1} + \cdots + \widehat{\zeta}_{k-l} x_k\}, \tag{7.21}$$

and the estimated dispersion parameter is $\widehat{\phi}$.

7.4.3 Interpretation of Estimated Regression Coefficients

It can be shown (do it!) that if π_0 and π_1 are fixed, then μ is the conditional expectation of y given $0 < y < 1$, i.e., $\mu = \mathbb{E}(y|0 < y < 1)$. Therefore,

- if x_1 is a continuous predictor, then $(\exp\{\widehat{\beta}_1\} - 1) \cdot 100\%$ gives the percent change in the estimated ratio $\dfrac{\widehat{\mu}}{1 - \widehat{\mu}} = \dfrac{\widehat{\mathbb{E}}(y|0 < y < 1)}{1 - \widehat{\mathbb{E}}(y|0 < y < 1)}$ for a one-unit increase in x_1, provided all the other predictors stay unchanged.

- If x_1 is an indicator variable, then $\exp\{\widehat{\beta}_1\} \cdot 100\%$ represents the percent ratio of $\dfrac{\widehat{\mu}}{1 - \widehat{\mu}}$ for $x_1 = 1$ and that for $x_1 = 0$.

Further, the parameters for this model are chosen in such a way (prove it!) that $\nu = \dfrac{\pi_0}{1 - \pi_0 - \pi_1} = \dfrac{\mathbb{P}(y = 0)}{\mathbb{P}(0 < y < 1)}$ and $\tau = \dfrac{\pi_1}{1 - \pi_0 - \pi_1} = \dfrac{\mathbb{P}(y = 1)}{\mathbb{P}(0 < y < 1)}$.
That means that ν represents the odds in favor of $y = 0$ versus $0 < y < 1$, and τ represents the odds of $y = 1$ against $0 < y < 1$. Consequently, estimated gamma coefficients are interpreted as in the logistic regression (see Subsection 3.1.3) in terms of odds of observing $y = 0$ versus $0 < y < 1$, and estimated zeta coefficients are interpreted as in the logistic regression in terms of odds of observing $y = 1$ against $0 < y < 1$.

7.4.4 Predicted Response

The mean of y in this model is equal to (show it!)

$$\mathbb{E}(y) = \pi_1 + (1 - \pi_0 - \pi_1)\,\mu = \frac{\tau + \mu}{1 + \nu + \tau},$$

and thus, in view of (7.19)-(7.21), the predicted response for given values of x variables, $x_1^0, x_2^0, \ldots, x_k^0$, may be found through the formula

$$y^0 = \left(\exp\{\widehat{\zeta}_0 + \widehat{\zeta}_1\,x_{l+1}^0 + \cdots + \widehat{\zeta}_{k-l}\,x_k^0\} + \frac{\exp\{\widehat{\beta}_0 + \widehat{\beta}_1\,x_1^0 + \cdots + \widehat{\beta}_m\,x_m^0\}}{1 + \exp\{\widehat{\beta}_0 + \widehat{\beta}_1\,x_1^0 + \cdots + \widehat{\beta}_m\,x_m^0\}} \right) \times$$

$$\times \left(1 + \exp\{\widehat{\gamma}_0 + \widehat{\gamma}_1\,x_{m+1}^0 + \cdots + \widehat{\gamma}_{l-m}\,x_l^0\} + \exp\{\widehat{\zeta}_0 + \widehat{\zeta}_1\,x_{l+1}^0 + \cdots + \widehat{\zeta}_{k-l}\,x_k^0\} \right)^{-1}.$$

7.4.5 SAS Implementation

Zero-one-inflated beta regression model can be programmed into SAS, using the `nlmixed` procedure. The general syntax is:

```
proc nlmixed data=data_name;
    parms b0=init_value <...> bm=init_value g0=init_value <...>
    g{1-m}=init_value z0=init_value <...> z{k-1}=init_value phi=init_value;
mu = exp(b0+b1*x1_name+···+bm*xm_name)/(1+exp(b0+b1*x1_name
    +···+bm*xm_name));
    nu = exp(g0+g1*x{m+1}_name+···+g{1-m}*xl_name);
        tau = exp(z0+z1*x{l+1}_name+z2*···+z{k-1}*xk_name);
    pi0=nu/(1+nu+tau);
        pi1=tau/(1+nu+tau);
    if (response_name=0) then loglikelihood_name=log(pi0);
        if (response_name=1) then loglikelihood_name=log(pi1);
    if (response_name >0 and response_name <1) then
        loglikelihood_name=log(1-pi0-pi1)+lgamma(phi)-lgamma(mu*phi)
            -lgamma((1-mu)*phi)+(mu*phi-1)*log(response_name)
            +((1-mu)*phi-1)*log(1-response_name);
model response_name~ general(loglikelihood_name);
        predict pi1+(1-pi0-pi1)*mu out=outdat_name;
run;


proc print data=outdata;
run;
```

7.4.6 R Implementation

In R, zero-one-inflated beta regression is fitted via function `gamlss()` in the library `gamlss`. The syntax is

```
fitted.model.name<- gamlss(response.name~x1.name+···+xm.name,
mu.link="logit", nu.formula=~x{m+1}.name+···+xl.name, nu.link="log",
tau.formula=~x{l+1}.name+···+xk.name, tau.link="log",
data =data.name, family=BEINF)

parampred.name<- predictAll(fitted.model,
newdata=data.frame(catpredictor1.name ="value", ...,
numpredictork.name=value), type="response")

print((parampred.name$tau+parampred.name$mu)/(1+parampred.name$nu
    +parampred.name$tau))
```

- In this model, the parameter Sigma is related to ϕ as $\text{Sigma}^2 = \dfrac{1}{1+\phi}$. Moreover, R outputs the value of (Intercept) that relates to Sigma via a logit link function. Thus, to recover the estimate of ϕ from (Intercept), we use the formula

$$\widehat{\phi} = \frac{1 - \text{Sigma}^2}{\text{Sigma}^2} = \frac{1 - \Big(\exp\{(\texttt{Intercept})\}/(1+\exp\{(\texttt{Intercept})\})\Big)^2}{\Big(\exp\{(\texttt{Iintercept})\}/(1+\exp\{(\texttt{Iintercept})\})\Big)^2}.$$

- Variables listed in newdata must be contained in the original data set *data_name*.

7.4.7   Example

EXAMPLE 7.4. As an experiment, a school district has implemented an online math coaching program in a single K-8 school. Administrators are interested in finding out what factors predict completion of weekly assignments. The data are collected on students' grade level (5, 6, 7, or 8, fitted as a continuous variable), gender (M/F), score in math class (in percent, can exceed 100 due to extra credit), and proportion of solved exercises in a certain weekly online assignment. Since some students may not have completed any exercises, a zero is a possible response. On the other hand, some students might have completed all of the exercises, so a one is also a plausible response. A proportion of completed exercises for all other students would be between zero and one. The setting calls for a zero-one-inflated beta regression, which is fitted below. Here, the parameter μ is regressed on gender, ν on mathscore, and τ on grade.


```
data assignment;
input grade gender$ mathscore propassign @@;
female=(gender='F');
cards;
5 M 87.5  0.62  7 F 68.4   0.00  8 F 100.0 0.95  6 M 84.5  0.50
7 F 76.1  1.00  5 M 87.3   0.32  6 M 66.9  0.80  5 F 91.5  1.00
7 M 90.6  0.45  6 M 85.9   0.27  7 M 97.1  0.55  6 F 104.9 0.85
8 F 98.2  1.00  5 F 103.8  0.95  5 F 80.2  0.97  8 M 56.1  0.00
7 M 68.1  0.00  7 M 76.3   0.70  6 M 97.3  0.30  8 F 66.3  1.00
7 M 79.0  0.00  6 M 73.9   0.46  8 M 77.3  0.22  7 M 80.2  0.72
7 M 77.1  1.00  8 F 103.1  1.00  7 M 104.8 1.00  7 F 55.7  0.00
7 M 69.4  0.67  5 F 83.6   0.68  8 F 81.4  1.00  7 M 84.7  0.59
```

```
6 F 92.1   0.51   5 M 97.8   0.65   7 M 88.2   0.00   8 F 69.5   0.96
6 M 94.8   0.21   7 M 84.7   0.35   5 M 67.7   0.00   7 F 100.3 1.00
8 M 104.2 0.92   7 F 77.2   0.20   7 M 70.9   0.73   6 M 96.7   0.62
;

proc nlmixed;
 parms b0=0.1 b1=0.1 g0=0.1 g1=0.1 z0=0.1 z1=0.1 phi=0.1;
  mu = exp(b0+b1*female)/(1+exp(b0+b1*female));
   nu = exp(g0+g1*mathscore);
    tau = exp(z0+z1*grade);
      pi0=nu/(1+nu+tau);
       pi1=tau/(1+nu+tau);
   if (propassign=0) then loglikelihood=log(pi0);
  if (propassign=1) then loglikelihood=log(pi1);
   if (propassign >0 and propassign <1) then
  loglikelihood =log(1-pi0-pi1)+lgamma(phi)-lgamma(mu*phi)
     -lgamma((1-mu)*phi)+(mu*phi-1)*log(propassign)
        +((1-mu)*phi-1)*log(1-propassign);
model propassign~general(loglikelihood);
run;
```

2 Log Likelihood 47.8

Parameter Estimates

Parameter	Estimate	Pr > |t|
b0	1.2472	0.0007
b1	1.1173	0.0064
g0	10.4743	0.0159
g1	-0.1532	0.0098
z0	-7.8921	0.0280
z1	0.9879	0.0491
phi	4.1694	

```
proc nlmixed;
 parms b0=0.1 g0=0.1 z0=0.1 phi=0.1;
   mu = exp(b0)/(1+exp(b0));
     nu = exp(g0);
        tau = exp(z0);
     pi0=nu/(1+nu+tau);
       pi1=tau/(1+nu+tau);
  if (propassign=0) then loglikelihood=log(pi0);
   if (propassign=1) then loglikelihood=log(pi1);
     if (propassign >0 and propassign <1) then
```

```
loglikelihood =log(1-pi0-pi1)+lgamma(phi)-lgamma(mu*phi)
    -lgamma((1-mu)*phi)+(mu*phi-1)*log(propassign)
    +((1-mu)*phi-1)*log(1-propassign);
model propassign~general(loglikelihood);
run;
```

2 Log Likelihood 73.3

```
data deviance_test;
 deviance=73.3-47.8;
   pvalue=1-probchi(deviance,3);
run;
```

```
proc print;
run;
```

deviance   pvalue
  25.5 0.000012136

The estimated parameters of the fitted model are

$$\widehat{\mu} = \frac{\exp\{1.2472 + 1.1173 * female\}}{1 + \exp\{1.2472 + 1.1173 * female\}}, \quad \widehat{\pi}_0 = \frac{\widehat{\nu}}{1 + \widehat{\nu} + \widehat{\tau}}, \quad \text{and} \quad \widehat{\pi}_1 = \frac{\widehat{\tau}}{1 + \widehat{\nu} + \widehat{\tau}},$$

where

$$\widehat{\nu} = \frac{\widehat{\mathbb{P}}(y = 0)}{\widehat{\mathbb{P}}(0 < y < 1)} = \exp\{10.4743 - 0.1532 * mathscore\},$$

and

$$\widehat{\tau} = \frac{\widehat{\mathbb{P}}(y = 1)}{\widehat{\mathbb{P}}(0 < y < 1)} = \exp\{-7.8921 + 0.9879 * grade\},$$

and $\widehat{\phi} = 4.1694$. This model fits the data well since the p-value in the goodness-of-fit test is way below 0.05. All the three predictors are significant at the 5% level.

For females, the estimated ratio $\dfrac{\widehat{\mu}}{1 - \widehat{\mu}} = \dfrac{\widehat{\mathbb{E}}(y|0 < y < 1)}{1 - \widehat{\mathbb{E}}(y|0 < y < 1)}$ is $\exp\{1.1173\} \cdot 100\% = 305.66\%$ of that for males. For a one-point increase in math score, the estimated odds in favor of $y = 0$ against $0 < y < 1$ change by $(\exp\{-0.1532\} - 1) \cdot 100\% = -14.20\%$, that is, decrease by 14.20%. As grade increases by one, the estimated odds in favor of $y = 1$ against $0 < y < 1$ increase by $(\exp\{0.9879\} - 1) \cdot 100\% = 168.56\%$.

Now we estimate the proportion of completed exercises for an 8th grade boy with 101% in math class. To this end, we compute

$$y^0 = \left(\exp\{-7.8921 + 0.9879 * 8\} + \frac{\exp\{1.2472\}}{1 + \exp\{1.2472\}} \right) \times$$

$$\times \left(1 + \exp\{10.4743 - 0.1532 * 101\} + \exp\{-7.8921 + 0.9879 * 8\} \right)^{-1} = 0.8861.$$

The SAS code for the prediction follows.

```
data prediction;
input grade mathscore female;
cards;
8 101 0
;

data assignment;
 set assignment prediction;
run;

proc nlmixed;
 parms b0=0.1 b1=0.1 g0=0.1 g1=0.1 z0=0.1 z1=0.1 phi=0.1;
   mu = exp(b0+b1*female)/(1+exp(b0+b1*female));
    nu = exp(g0+g1*mathscore);
      tau = exp(z0+z1*grade);
    pi0=nu/(1+nu+tau);
     pi1=tau/(1+nu+tau);
       if (propassign=0) then loglikelihood=log(pi0);
   if (propassign=1) then loglikelihood=log(pi1);
      if (propassign >0 and propassign <1) then
   loglikelihood =log(1-pi0-pi1)+lgamma(phi)
   -lgamma(mu*phi)-lgamma((1-mu)*phi)+
      (mu*phi-1)*log(propassign)+((1-mu)*phi-1)*log(1-propassign);
  model propassign~general(loglikelihood);
     predict (tau+mu)/(1+nu+tau) out=outdata;
run;

proc print data=outdata (firstobs=45 obs=45);
var Pred;
run;


   Pred
0.76493
```

The R script and output are below.

```
assignment.data<-read.csv(file="./Example7.4Data.csv",
header=TRUE, sep=",")
install.packages("gamlss")
library(gamlss)

#computing new variable and making it part of the original
  #data set
assignment.data$female<- ifelse(assignment.data$gender=="F",1,0)

#fitting zero-one-inflated beta regression model
summary(fitted.model<- gamlss(propassign~ female,
mu.link="logit", nu.formula=~ mathscore, nu.link="log",
tau.formula=~ grade, tau.link="log",
data=assignment.data, family=BEINF))
```

Mu link function: logit
Mu Coefficients:

	Estimate	Pr(>|t|)
(Intercept)	1.2467	0.000782
female	1.1169	0.006916

Sigma link function: logit
Sigma Coefficients:

	Estimate
(Intercept)	-0.2418

Nu link function: log
Nu Coefficients:

	Estimate	Pr(>|t|)
(Intercept)	10.47368	0.0163
mathscore	-0.15321	0.0101

Tau link function: log
Tau Coefficients:

	Estimate	Pr(>|t|)
(Intercept)	-7.8920	0.0290
grade	0.9879	0.0502

```
#checking model fit
intercept.only.model<- gamlss(propassign~ 1, mu.link="logit",
```

```
nu.formula=~ 1, nu.link="log", tau.formula=~ 1, tau.link="log",
data=assignment.data, family=BEINF)
print(deviance<- -2*(logLik(intercept.only.model)
-logLik(fitted.model)))
```

25.51013

```
print(p.value<- pchisq(deviance, df=3, lower.tail=FALSE))
```

1.207707e-05

```
#using fitted model for prediction
param.pred<- predictAll(fitted.model, newdata
=data.frame(grade=8, mathscore=101, female=0), type="response")
print((param.pred$tau+param.pred$mu)/
  (1+param.pred$nu+param.pred$tau))
```

0.7649165

□

Exercises for Chapter 7

EXERCISE 7.1. Prove that for the beta regression defined by (7.1) and (7.2), the mean of y is $\mathbb{E}(y) = \mu$, and the variance of y is $\mathbb{V}ar(y) = \dfrac{\mu(1-\mu)}{1+\phi}$.

EXERCISE 7.2. Ornithologists have collected data on migration of birds. They ringed 19 flocks of migratory birds prior to migration, and recorded for each flock the number of ringed birds, average mass (in grams), and average wingspan (in cm). The ringed flocks were observed later at the wintering grounds and the number of successfully migrated birds were counted. The distances traveled (in km) were also observed. The data are presented below.

Mass	Wingspan	Distance	N birds ringed	N birds migrated
811	67	1680	70	8
261	33	2137	113	75
398	48	2159	100	51
114	56	1204	145	113
119	53	1673	72	28
151	30	543	87	71
176	70	1414	116	109
184	45	2296	90	68
250	42	1511	52	42
505	24	741	74	63
551	17	1434	114	105
716	51	2116	98	58
735	119	2171	98	35
1233	108	2442	69	13
1315	98	2061	61	38
1633	72	1955	81	24
1736	119	1297	71	70
2019	101	930	112	105
2476	100	2312	95	37

(a) Model the proportion of birds per flock that successfully reach the winter grounds. To avoid small estimates of the regression coefficients, convert mass into kilograms, wingspan into meters, and the distance, into thousands of kilometers. Write out the fitted model explicitly. Which predictors are significant at the 5% level?

(b) Analyze the model fit.

(c) Give interpretation of the estimated significant parameters.

(d) Predict the proportion of birds that successfully reach the winter grounds for a flock of birds with average mass of 600 grams, average wingspan of 65 centimeters, that travel a distance of 1650 kilometers, and typically travel in flocks of about 70 birds.

EXERCISE 7.3. The department of Health Care Administration is analyzing utilization of hospital resources. They retrieve data on 30 randomly chosen U.S. hospitals, and are looking at percent of emergency room (ER) patients who were hospitalized, hospital location (urban or rural), hospital type (public or private), and hospital size by number of beds. The data are:

Percent hospzd	Hosp loc	Hosp type	N beds	Percent hospzd	Hosp loc	Hosp type	N beds
17	rural	private	56	4	urban	public	91
39	rural	public	144	6	urban	public	77
38	urban	public	61	39	urban	private	237
48	rural	public	186	41	urban	private	56
30	rural	private	132	45	rural	public	43
25	urban	private	589	13	urban	public	64
5	urban	public	53	42	rural	public	193
4	rural	private	73	28	urban	private	363
48	rural	private	154	31	urban	public	600
4	urban	public	38	48	rural	public	468
26	rural	private	318	41	rural	public	311
15	urban	public	35	9	urban	public	65
28	urban	private	184	13	urban	private	44
34	urban	private	173	44	urban	public	479
31	urban	public	63	16	rural	public	72

(a) Model the proportion of hospitalized ER patients. Write down the fitted model. What factors are significant predictors? Use $\alpha = 0.05$.

(b) How good is the model fit?

(c) Interpret estimated significant regression coefficients.

(d) Give the predicted proportion of hospitalized ER patients for a rural public hospital with 50 beds.

EXERCISE 7.4. In commercial fishing, an issue of concern is large percent of by-catch, an unwanted marine life caught along with intended fish. An analyst at a tuna fishery collected data on a randomly chosen sample of catches made by 20 fishing vessels. The variables are the distance to shore (in nautical miles), method of fishing (trawling, purse seining, or longline fishing), fishing depth (in meters), and the percentage of by-catch. The data are:

Distance to shore	Fishing method	Fishing depth	Percent by-catch	Distance to shore	Fishing method	Fishing depth	Percent by-catch
120	trawl	250	14	115	trawl	300	8
115	trawl	150	6	160	trawl	200	10
70	trawl	300	24	160	trawl	200	10
130	trawl	150	6	50	trawl	150	15
90	seine	200	56	10	seine	150	16
15	seine	350	32	25	seine	200	22
15	seine	150	13	15	seine	300	21
20	seine	350	23	40	longline	100	21
15	longline	200	10	60	longline	200	4
40	longline	150	7	50	longline	150	17

(a) Use the beta regression to model proportion of by-catch. Convert depth to kilometers. Write down the fitted model.

(b) Discuss significance of predictor variables and model fit.

(c) Give interpretation of the estimates of the regression coefficients for the significant predictors.

(d) Find the predicted percent of by-catch for a trawler that fishes 80 nautical miles off shore at the depth of 250 meters.

EXERCISE 7.5. A real estate intern is interested in modeling the proportion of houses that were sold in one month as having beta distribution with predictor variables: an average house price in subdivision (in thousands of dollars), the number of houses in subdivision, and age of subdivision (in years). The data for the past year on 30 randomly chosen subdivisions were available. They are:

Percent sold	Average price	N houses	Age	Percent sold	Average price	N houses	Age
0	455	69	21	80	308	223	15
0	316	244	24	75	159	84	13
36.4	210	236	31	0	147	54	37
50	557	183	16	44.4	704	199	18
33.3	232	73	6	0	593	119	38
50	626	230	20	20	738	156	8
27.3	343	60	14	55.6	256	206	34
80	246	201	17	85.7	345	38	22
42.9	631	217	11	50	450	158	7
0	630	222	42	0	491	239	27
71.4	356	85	22	28.6	441	103	15
25	481	240	16	88.9	212	222	18
50	181	197	42	50	574	56	35
20	264	235	19	33.3	647	138	35
87.5	297	88	17	0	630	18	60

(a) Fit the zero-inflated beta regression model to the proportion of houses. Regress the probability of zero on age of subdivision. To achieve model convergence, normalize the average price and number of houses by a factor of 100. Discuss significance of the predictors at the 5% level.

(b) Present the fitted model. Does this model have a decent fit?

(c) Interpret parameter estimates for statistically significant predictors.

(d) What is the model prediction for percent houses sold for a subdivision with 300 houses, built 50 years ago, and where houses are sold, on average, for $450,000?

EXERCISE 7.6. Head instructors at a chain of traditional martial arts studios want to see if there is an association between the proportion of first-place trophies a studio wins during a tournament and such predictors as the number of years a studio has been in existence, the number of pupils, and the number of black-belt instructors. For each studio, the proportion of first-place trophies is defined as the ratio of the number of first-place trophies to the total number of trophies a studio wins. Data are collected on 26 studios.

Trophies	Firstplaces	Years	Blackbelts	Pupils
21	7	5	1	96
12	3	5	2	59
21	10	5	2	71
23	4	3	2	94
11	1	1	3	53
20	9	6	4	52
15	4	6	2	61
28	16	13	5	104
19	8	3	4	95
4	0	1	1	77
6	0	1	1	45
19	12	7	5	42
21	7	4	3	86
32	24	11	6	151
5	0	3	1	58
23	9	5	2	81
8	0	3	2	45
21	13	15	3	89
12	3	6	3	39
11	0	3	2	40
12	7	5	2	81
22	13	7	4	148
10	3	8	3	128
20	2	2	2	42
19	2	3	1	39
14	2	2	3	105

(a) Model the proportion of first-place trophies using a zero-inflated beta regression. Use the number of pupils to predict the probability of zero. Specify the fitted model. Use alpha of 0.05 to determine significance of regression coefficients.

(b) Discuss the model fit. Present the fitted model.

(c) Interpret the estimates of significant regression coefficients.

(d) Predict the proportion of first-place trophies won by a studio that has been around for 10 years, has 85 students and three black-belt instructors.

EXERCISE 7.7. Parks and Recreation Department is conducting a study on mortality of young trees planted in parks. Investigators have collected data on the number of planted trees, number of trees that survived for two years, frequencies of pest control and soil fertilization, average annual precipitation (in inches), and average annual wind speed (in miles per hour). The data are as follows:

N planted trees	N survived trees	Freq of pest control	Freq of fertilization	Precip	Wind speed
125	125	3	1	18	9.6
115	68	0	0	8	13.4
250	101	1	1	17	12.8
95	85	2	2	22	10
140	48	3	1	15	15.1
75	75	3	2	27	6.3
185	163	3	3	15	12.3
20	9	3	0	18	9.4
110	83	3	1	24	13.1
80	80	0	1	18	7.8
120	117	4	1	20	9.3
90	56	5	1	15	13.9
30	30	3	0	33	8.6
90	81	4	1	23	7.7
140	119	3	1	18	11.8
70	9	3	0	32	8.4
75	71	3	3	20	13.4
150	102	5	0	16	9.7
90	73	4	1	15	9.7
160	151	6	1	18	7.8
100	46	3	1	20	12.3
85	85	4	1	22	6.8
120	85	2	1	19	6.6
180	53	3	1	29	9.4
45	12	0	1	9	13.1
35	35	1	0	7	12.4

(a) Fit a one-inflated beta regression to model the proportion of survived trees. Regress the probability of one on amount of precipitation and wind speed. Discuss significance of the predictors at 5% and 10% significance levels.

(b) Present the fitted model and discuss its fit.

(c) Interpret the estimated significant parameters.

(d) Parks and Recreation Department employees are considering planting 100 trees in a hard to reach area where neither pest control nor soil fertilization would be feasible. They are trying to decide between an area with lower precipitation (2 inches) and stronger winds (12.5mph), and an area with higher precipitation (25 in) and lower winds (6 mph). Which of the two areas would you recommend to use based on predicted proportion of trees that would survive for two years?

EXERCISE 7.8. A manager of a store that specializes in selling name-brand travel luggage is interested in finding out what characteristics of a salesperson best predict the proportion of initiated sales that are successfully completed. He looks up records of the sales team members and collects the data on gender, years of experience, amount of bonus received the previous year and the proportion of completed sales. The data are given in the table below.

Gender	Expyr	Bonus	Propsales	Gender	Expyr	Bonus	Propsales
F	1	1.1	0.67	F	2	1.2	0.65
M	11	0.5	1.00	F	13	0.6	1.00
M	4	1.1	0.90	F	8	0.9	0.54
M	2	1.6	0.93	M	4	0.6	0.63
F	2	0.7	0.49	F	17	2.4	1.00
F	4	1.05	0.88	F	3	1.6	1.00
M	1	1.6	0.96	F	2	1.4	0.88
F	2	1.2	0.67	F	4	1.05	0.85
M	2	1.6	0.94	F	8	1.4	1.00
M	7	1.4	0.77	M	4	1.35	0.95
F	4	1.55	1.00	F	3	1	0.83
F	4	0.9	0.51	F	18	1.25	1.00
F	8	0.95	0.59	M	4	0.4	0.66

(a) Fit a one-inflated beta regression to model the proportion of completed sales, regressing the probability of one on the number of years of experience a salesperson has accrued. Use the significance level of 0.05. Write down the fitted model.
(b) How good is the model fit?
(c) Interpret the estimated significant regression coefficients.
(d) Predict the proportion of completed sales for a salesman with 3 years of work experience and who received $1,500 in bonuses the previous year.

EXERCISE 7.9. An agricultural laboratory is testing germination of corn seeds at various locations. The variables of interest are soil electrical conductivity (EC, in millisiemens per centimeter squared), mean soil temperature (in degrees Fahrenheit), plot altitude (in feet), and rate of seed germination. At some locations seeds didn't germinate at all, whereas in some other locations they germinated at 100% rate. The data are provided below.

EC	Soiltemp	Altitude	Germrate	EC	Soiltemp	Altitude	Germrate
2.7	67	4368	0.00	2.5	64	4229	0.17
1.1	67	1689	1.00	1.8	69	2933	0.47
1.8	69	3156	0.87	1.8	63	6110	0.32
1.6	67	4884	0.58	1.5	67	461	1.00
2.4	66	4926	0.00	2.5	67	5269	0.00
1.6	63	3854	0.23	1.7	74	197	1.00
2.3	67	5146	0.00	1.6	65	607	1.00
1.2	64	2202	0.48	2.6	67	5263	0.16
1.1	62	2759	0.82	1.2	69	651	1.00
1.9	62	2774	0.61	1.7	65	863	0.80
1.5	71	5927	0.19	1.5	62	4386	0.23
1.7	61	827	0.93	1.7	68	165	1.00
2.8	62	3631	0.00	1.7	62	234	0.73

(a) Fit the beta regression with inflated zeros and ones to model the germination rate. Regress parameter μ on altitude normalized by a factor of 1000, ν on EC, and τ on soil temperature. Specify the fitted model. Determine significance of regression coefficients at 5% and 10% levels.

(b) Discuss the model fit.

(c) Give interpretation of the estimated significant regression coefficients.

(d) Use the fitted model to predict the germination rate for a plot with EC of 1.5 mS/cm^2, soil temperature of 68°F, and altitude of 950 feet.

EXERCISE 7.10. A college football analytic team is trying to determine what attributes of a player predict the proportion or games played in a season by that player. They collect data retrospectively on BMI, forty yard dash (in seconds), vertical jump (in inches), broad jump (in inches), number of bench press reps at 225 pounds, and proportion of games played. Some players went through the training but never played in big games, while some others were very successful and played in every game. The data are given below.

BMI	Fortyyd	Vertical	Broad	Bench	Propgames
31.1	4.58	35.0	108	20	0.00
28.5	4.70	30.5	115	21	1.00
32.4	4.39	36.0	116	18	0.00
30.8	4.67	33.0	121	15	0.87
29.6	4.41	33.0	116	26	0.47
30.0	4.56	38.0	122	21	0.80
31.3	4.50	35.0	119	29	0.80
29.4	4.49	31.5	115	18	0.33
28.1	4.37	34.5	130	21	0.53
31.0	4.52	33.5	128	25	0.87
29.7	4.57	31.5	124	18	0.73
27.6	4.62	38.5	118	15	1.00
29.5	4.60	32.0	121	15	0.47
30.7	4.40	34.5	113	22	0.47
29.4	4.73	34.0	114	15	0.60
29.1	4.57	35.0	111	19	0.87
30.6	4.60	30.5	114	24	0.47
29.3	4.55	36.0	115	17	0.87
28.1	4.59	35.5	109	14	0.87
31.7	4.62	37.0	121	19	0.80
29.7	4.73	34.0	118	21	1.00
30.7	4.80	34.0	114	15	0.33
28.8	4.37	36.0	121	19	0.73
30.7	4.68	34.0	105	14	0.00
30.7	4.64	33.0	116	21	0.47
30.3	4.50	35.5	115	25	0.60
30.0	4.48	34.0	119	20	0.67
28.6	4.59	36.0	123	17	1.00
30.8	4.43	34.0	117	18	0.53
27.7	4.51	33.0	117	20	0.73
30.7	4.50	38.0	114	18	1.00
28.1	4.59	36.0	115	17	0.93
29.3	4.61	32.0	113	20	0.00
27.9	4.64	33.0	118	23	1.00
29.7	4.67	41.0	124	21	1.00
29.9	4.48	35.0	111	18	0.00
32.0	4.51	33.5	116	25	0.67
29.6	4.37	37.0	115	24	0.93
32.9	4.55	33.0	122	27	0.47
27.6	4.67	33.5	118	26	1.00
30.4	4.55	32.0	109	17	0.73
31.4	4.50	38.0	121	18	0.67

(a) Regress the proportion of games played, using the zero-one-inflated beta model. Regress μ on vertical jump and number of bench press reps, ν on broad jump, and τ on BMI and forty yard dash. What predictors are significant at the 5% level?

(b) Analyze the fit of the model.

(c) Give interpretation of the estimated significant coefficients.

(d) Predict the proportion of games that a new player will play, if his BMI is 27.8 kg/m^2, forty dash run is 4.67 seconds, vertical jump is 32 inches, broad jump is 117 inches, and bench press is 16 reps.

Chapter 8

General Linear Regression Models for Repeated Measures Data

In the previous chapters, observations for different individuals were assumed independent. In this chapter, we will consider cases when multiple observations are collected on the same individuals repeatedly over time, or space, or under different treatment conditions. Such observations are referred to as *repeated measures*. In these models, observations for different individuals are assumed uncorrelated. However, observations for each individual are modeled as correlated. This potential correlation is included into the model by introducing additive random terms. Thus, along with usual predictors (termed *fixed-effects predictors*), the model also contains random variables (called *random-effects terms*). Such a model is generally referred to as a *mixed-effects linear regression model*[1] (or, simply, a *mixed* model).

A special case of repeated measures is *longitudinal data*, which are data collected at several time points. Below we will give the theoretical framework referring to longitudinal data and using variable *time* to index repeated observations. The specificity of longitudinal data is that times between the repeated measures are not necessarily equally distant.

8.1 Random Slope and Intercept Model

8.1.1 Model Definition

Suppose data are collected longitudinally at times $t_1, \ldots, t_p$. For each individual i, $i = 1, \ldots, n$, at time t_j, $j = 1, \ldots, p$, the observations are $x_{1ij}, \ldots, x_{kij}, y_{ij}$.

---

[1]Mixed-effect modeling was introduced in Laird, Nan M. and James H. Ware (1982). "Random-Effects Models for Longitudinal Data". *Biometrics*, 38(4): 963 – 974.

The *random slope and intercept model* is defined as

$$y_{ij} = \beta_0 + \beta_1 x_{1ij} + \cdots + \beta_k x_{kij} + \beta_{k+1} t_j + u_{1i} + u_{2i} t_j + \varepsilon_{ij} \qquad (8.1)$$

where u_{1i}'s are independent $\mathcal{N}(0, \sigma_{u_1}^2)$ *random intercepts*, u_{2i}'s are independent $\mathcal{N}(0, \sigma_{u_2}^2)$ *random slopes*, and ε_{ij}'s are independent $\mathcal{N}(0, \sigma^2)$ *errors* that are also independent of u_{1i}'s and u_{2i}'s. It is assumed that $\mathbb{C}ov(u_{1i}, u_{2i}) = \sigma_{u_1 u_2}$, and $\mathbb{C}ov(u_{1i}, u_{2i'}) = 0$ for $i \neq i'$.

Note that in this model the predictor variables $x_1, \ldots, x_k$ have fixed effects, whereas the random intercept u_1 and slope u_2 have random effects. It is also noteworthy that the predictors $x_1, \ldots, x_k$ may vary with time.

It can be shown (see Exercise 8.1) that $\mathbb{C}ov(y_{ij}, y_{i'j'}) = 0$, $i \neq i'$, meaning that the responses for different individuals are uncorrelated for any time points. It can also be shown that responses between different time points for the same individual may be correlated, since $\mathbb{C}ov(y_{ij}, y_{ij'}) = \sigma_{u_1}^2 + \sigma_{u_1 u_2}(t_j + t_{j'}) + \sigma_{u_2}^2 t_j t_{j'}$, for $j \neq j'$. In addition, it can be verified that the response variable y_{ij} is normally distributed with the mean

$$\mathbb{E}(y) = \beta_0 + \beta_1 x_1 + \cdots + \beta_k x_k + \beta_{k+1} t \qquad (8.2)$$

and variance $\mathbb{V}ar(y_{ij}) = \sigma_{u_1}^2 + 2\sigma_{u_1 u_2} t_j + \sigma_{u_2}^2 t_j^2 + \sigma^2$.

8.1.2   Fitted Model

By (8.2), in the random slope and intercept model, the fitted mean response can be expressed as

$$\widehat{\mathbb{E}}(y) = \widehat{\beta}_0 + \widehat{\beta}_1 x_1 + \cdots + \widehat{\beta}_k x_k + \widehat{\beta}_{k+1} t. \qquad (8.3)$$

All beta parameters along with $\sigma_{u_1}^2, \sigma_{u_1 u_2}, \sigma_{u_2}^2$, and σ^2 are estimated by maximizing the likelihood function, or, alternatively, by maximizing the restricted likelihood function. The *restricted maximum likelihood* (REML) method utilizes a certain linear transformation of y, chosen in such a way that the resulting likelihood function doesn't depend on β's. It is known that the maximum likelihood estimators of the variances and covariance are biased, whereas the REML method produces unbiased estimators.

8.1.3   Interpretation of Estimated Regression Coefficients

In view of (8.3), the fitted beta coefficients are interpreted as in the general linear regression model (see Subsection 1.5), in terms of an average change in the mean response for a unit-increase in a continuous predictor, or as the difference between mean responses for level 1 and level 0 for a 0-1 predictor, provided all the other predictors stay unchanged.

8.1.4 Model Goodness-of-Fit Check

To compare the fit of several models, the AIC, AICC, and BIC criteria may be used.

To check goodness-of-fit of a particular model, a chi-squared deviance test is used (see Subsection 1.6). In this case, the null model has only fixed-effects predictor variables. Since the random-effects terms bring three parameters into the model (two variances and a covariance), this test has three degrees of freedom.

8.1.5 Predicted Response

By (8.3), for a deterministic set of values $x_1^0, \ldots, x_k^0, t^0$, the predicted response is $y^0 = \widehat{\beta}_0 + \widehat{\beta}_1\, x_1^0 + \cdots + \widehat{\beta}_k\, x_k^0 + \widehat{\beta}_{k+1}\, t^0$.

8.1.6 SAS Implementation

Suppose the data set is given in the *short-and-wide form*, that is, it contains n rows for the n individuals: one column for individual's id, k columns for the variables $x_1, \ldots, x_k$, and p columns for the responses at the p time points $y_1, \ldots, y_p$. Prior to fitting a mixed-effects model, a *long-form data set* must be created, a data set that contains p rows for each individual, one row for each time point. This can be done via the following data statement:

```
data longform_data_name;
   set data_name;
     array time_array_name[p] (time1_name...timep_name);
     array response_array_name[p] y1_name...yp_name;
   do count_variable_name=1 to p;
    time_name=time_array_name[count_variable_name];
    response_name=response_array_name[count_variable_name];
     output;
   end;
   keep id_name x1_name...xk_name time_name response_name;
run;
```

The procedure `mixed` fits mixed-effects models on long-form data sets that contain one row per individual per time point. The general statements are:

```
proc mixed data=longform_data_name covtest;
   class <list of categorical predictors>;
      model response_name=<list of x predictors> time_name/solution
                             outpm=outdata_name;
         random intercept time_name/subject=id_name type=un;
run;
```

- The option `covetest` requests p-values in the tests for significance of the variance-covariance parameters.
- SAS outputs estimates of the beta coefficients only if the option `solution` is specified.
- The `outpm=` produces prediction of the response y for each row in the data set. If a new prediction is desired, the case must be added to the data set.
- Specification of `type=un`, an unstructured type of variance-covariance matrix, requests an estimate of the covariance $\sigma_{u_1 u_2}$.
- In the output, the estimators of $\sigma_{u_1}^2, \sigma_{u_1 u_2}, \sigma_{u_2}^2$, and σ^2 are termed `UN(1,1)`, `UN(2,1)`, `UN(2,2)` and `Residual`, respectively.

8.1.7   R Implementation

To create a long-form data set from a short-and-wide form *data.name* with columns *id.name*, *x1.name*, ..., *xk.name*, *y1.name*, ..., *yp.name*, the function `melt()` in the library `reshape2` may be used. The script is as follows:

```
install.packages("reshape2")
library(reshape2)
```

longform.data.name<- melt(*data.name*, id.vars=c("*id.name*", "*x1.name*", ..., "*xk.name*"), variable.name = "*response.variables.name*", value.name="*response.name*")

Here the categorical variable *response.variables.name* contains the names of the variables *y1.name*, ..., *yp.name*, whereas the variable *response.name* contains the actual numeric responses.

In a more general case, the short-and-wide data set might contain one predictor variable that varies with time, that is, the columns could be *id.name*, *x1.name*, ..., *x{k-1}.name*, *xk1.name*, ..., *xkp.name*, *y1.name*, ..., *yp.name*. A long-form data set then can be created by separating the response columns from the rest of the data, turning the two files into long form, and then merging them. The syntax is:

longform.data1.name<- melt(*data.name*[,c("*id.name*", "*x1.name*", ..., "*x{k-1}.name*", "*xk1.name*",..., "*xkp.name*")], id.vars=c("*id.name*", "*x1.name*", ..., "*x{k-1}.name*"), variable.name="*xk.variables.name*", value.name="*xk.name*")

longform.data2.name<- melt(*data.name*[,c("*y1.name*", ..., "*yp.name*")], variable.name="*response.variables.name*", value.name="*response.name*")

longform.data.name<- cbind(*longform.data1.name*, *longform.data2.name*)

If more than one predictor depends on time, then it would be necessary to split the original data set into several data sets, transform them into long form, and then merge together.

Next, the variable *response.variables.name* has to be recoded into a numeric variable that contains time values. This can be done as follows:

time.name<- ifelse(*longform.data.name$response.variables.name*=="*y1.name*", *time1.value*, ifelse(*longform.data.name$response.variables.name*=="*y2.name*", *time2.value*,..., ifelse(*longform.data.name$response.variables.name*== "*y{p-1}.name*", *time{p-1}.value*, *timep.value*)))

Next, for the long-form data set, the function lme() in the library nlme fits the random slope and intercept model. The statements are as follows:

summary(*fitted.model.name*<- lme(*response.name*~*x1.name*+··· +*xk.name* +*time.name*, random=~ 1+*time.name*|*id.name*, data=*data.name*)) intervals(*fitted.model.name*)

- R outputs estimated standard deviations and correlation coefficient for the random terms as opposed to SAS that outputs estimated variances and covariance.
- R doesn't conduct tests of significance for the variance-covariance parameters. The 95% confidence intervals based on a normal approximation may be requested by including the intervals() function.
- R outputs AIC and BIC values automatically. The AICC value has to be computed as

print(AICC<- -2*logLik(*fitted.model.name*)+2*p*n/(n-p-1))

- To obtain a predicted response value, use the function `predict()` with the added argument `level=0`.

8.1.8 Example

EXAMPLE 8.1. In a clinic, doctors are testing a certain cholesterol lowering medication. Patients' gender and age at the beginning of the study are recorded for 27 patients. The low-density lipoprotein (LDL) cholesterol levels are measured in all the patients at the baseline, and then at 6-, 9-, and 24-month visits. The SAS code below creates a long-form data set.

```
data cholesterol;
input id gender$ age LDL0 LDL6 LDL9 LDL24 @@;
cards;
1  M 50 73  71  80  85   2  F 72 174 164 139 112
3  M 46 85  86  82  90   4  F 71 172 150 139 127
5  F 75 186 177 153 145  6  F 68 184 169 153 138
7  F 63 196 188 163 155  8  M 73 137 137 132 104
9  M 59 135 120 106 106  10 M 60 111 110 100 76
11 F 59 127 126 106 99   12 M 46 88  87  84  80
13 F 67 176 150 156 153  14 F 52 155 135 128 120
15 M 65 142 117 114 97   16 F 75 158 143 145 135
17 F 57 148 131 138 102  18 M 58 125 111 118 124
19 M 48 76  65  94  98   20 M 47 116 108 94  107
21 F 53 191 185 162 113  22 F 73 167 165 162 140
23 M 62 109 104 93  94   24 F 77 167 164 155 155
25 M 55 103 94  75  78   26 F 74 122 126 105 111
27 F 79 203 204 178 145
;

data longform;
 set cholesterol;
   array m[4] (0 6 9 24);
  array c[4] LDL0 LDL6 LDL9 LDL24;
   do i=1 to 4;
   month=m[i];
        LDL=c[i];
     output;
   end;
keep id gender age month LDL;
run;
```

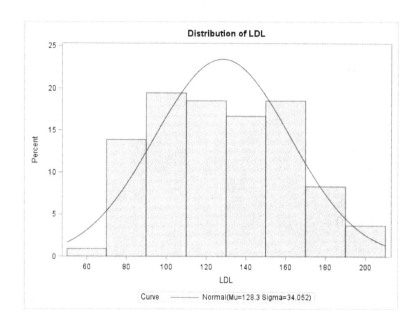

Figure 8.1: Histogram for LDL in SAS

Next, we show that the variable LDL, which contains all the cholesterol measurements, has a normal distribution. To this end, we construct a histogram with fitted bell-shaped curve and conduct normality testing.

```
proc univariate data=longform;
 var LDL;
  histogram LDL/normal;
run;
```

The output follows. Since the histogram resembles a normal density and all the p-values exceed 0.05, we conclude that the response variable is normally distributed.

Goodness-of-Fit Tests for Normal Distribution

Test	p Value
Kolmogorov-Smirnov	>0.150
Cramer-von Mises	0.126
Anderson-Darling	0.106

Our next step is to fit a random slope and intercept model, regressing LDL on the other variables. The code and relevant output are given below.

```
proc mixed covtest;
```

```
   class gender;
     model LDL=gender age month/solution;
       random intercept month/subject=id type=un;
run;
```

Covariance Parameter Estimates

Cov Parm	Estimate	Pr Z
UN(1,1)	520.17	0.0013
UN(2,1)	-16.3953	0.0087
UN(2,2)	0.7846	0.0028
Residual	69.8556	<.0001

Fit Statistics

AIC	870.3
AICC	870.7
BIC	875.5

Null Model Likelihood Ratio Test

DF	Chi-Square	Pr > ChiSq
3	72.86	<.0001

Solution for Fixed Effects

Effect	gender	Estimate	Pr > |t|
Intercept		65.0163	0.0029
gender	F	29.8105	<.0001
gender	M	0	.
age		0.9203	0.0085
month		-1.0957	<.0001

From the output, the variances of the random slope, intercept, and error, as well as the covariance term are all statistically significant, which validates the use of all the random terms in the model. The null-model likelihood test gives a small p-value, indicating a good fit. The fitted model can be written as $\widehat{E}(LDL) = 65.0163 + 29.8105 * female + 0.9203 * age - 1.0957 * month$, with all predictors being significant. For female patients, the estimated average cholesterol is 29.8105 points higher than that for males. With a one-year increase in age, the cholesterol level goes up, on average, by estimated 0.9203 points. As time increases by one month, the estimated average LDL measurement decreases by 1.0957 points.

The values for the AIC, AICC, and BIC criteria that are included in the above output will be used in Example 8.2.

To predict the LDL level at 3 months for a 48-year-old female patient, we compute $LDL^0 = 65.0163 + 29.8105 + 0.9203 * 48 - 1.0957 * 3 = 135.7141$.

To produce the same result in SAS, we run the following statements:

```
data prediction;
input id gender$ age month;
cards;
28 F 48 3
;

data longform;
 set longform prediction;
run;

proc mixed covtest;
class gender;
model LDL=gender age month/solution outpm=outdata;
random intercept month/subject=id type=un;
run;

proc print data=outdata (firstobs=109 obs=109);
var Pred;
run;

   Pred
135.716
```

The R script for this example is presented below.

```
cholesterol.data<-read.csv(file="./Example8.1Data.csv",
header=TRUE, sep=",")

#creating long-form data set
install.packages("reshape2")
library(reshape2)

longform.data<- melt(cholesterol.data,
id.vars=c("id", "gender", "age"),
variable.name="LDLmonth", value.name="LDL")
```

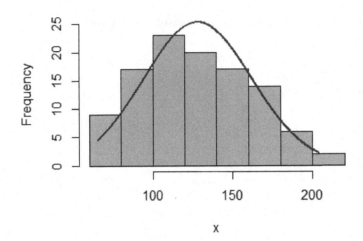

Figure 8.2: Histogram for LDL in R

```
#creating numeric variable for time
month<- ifelse(longform.data$LDLmonth=="LDL0", 0,
ifelse(longform.data$LDLmonth
=="LDL6", 6, ifelse(longform.data$LDLmonth=="LDL9",9,24)))

#plotting histogram with fitted normal density
install.packages("rcompanion")
library(rcompanion)

plotNormalHistogram(longform.data$LDL)

#testing for normality of distribution
shapiro.test(longform.data$LDL)
```

Shapiro-Wilk normality test
W = 0.97668, p-value = 0.05449

```
#specifying reference categories
gender.rel<- relevel(longform.data$gender, ref="M")

#fitting random slope and intercept model
install.packages("nlme")
library(nlme)
```

```
summary(fitted.model<- lme(LDL ~ gender.rel + age + month,
random=~ 1 + month|id, data=longform.data))
```

Linear mixed-effects model fit
```
      AIC       BIC      logLik
 878.3294  899.4845  -431.1647
```

Random effects:
```
              StdDev    Corr
(Intercept) 22.8072879  (Intr)
month        0.8857844 -0.812
Residual     8.3579458
```

Fixed effects:
```
              Value     p-value
(Intercept)  65.01614   0.0014
gender.relF  29.81056   0.0003
age           0.92033   0.0116
month        -1.09566   0.0000
```

```
intervals(fitted.model)
```

Random Effects:
```
                          lower        est.        upper
sd((Intercept))        16.4912656  22.8072879  31.542296
sd(month)               0.6229652   0.8857844   1.259483
cor((Intercept),month) -0.9325355  -0.8115565  -0.526760
```

  Within-group standard error:
```
    lower      est.      upper
 6.921510  8.357946  10.092488
```

```
#computing AICC
n<- 108
p<- 8
print(AICC<- -2*logLik(fitted.model)+2*p*n/(n-p-1))
```

879.784

```
#checking model fit
null.model<- glm(LDL ~ gender.rel + age + month,
data=longform.data)
print(deviance<- -2*(logLik(null.model)-logLik(fitted.model)))
```

77.80585

```
print(p.value<- pchisq(deviance, df=3, lower.tail=FALSE))
```

9.069804e-17

```
#using fitted model for prediction
print(predict(fitted.model, data.frame(gender.rel="F",
age=48, month=3), level=0))
```

135.7156

□

8.2 Random Slope and Intercept Model with Covariance Structure for Error

8.2.1 Model Definition

In the random slope and intercept model introduced in (8.1), the variance-covariance matrix for the error terms has a diagonal structure. It is of the form $\sigma^2 \mathbb{I}$, where $\mathbb{I}$ denotes a $np \times np$ identity matrix, with ones on the main diagonal and zeros everywhere else. This structure assumes that the errors are independent, and thus is termed an *independent structure*.

The random slope and intercept models with other structures for the variance-covariance matrix of the error terms can be considered. The most general is an *unstructured* one with an $np \times np$ block-diagonal matrix with n identical blocks of size $p \times p$ of the form:

$$
\begin{bmatrix}
\sigma_1^2 & \sigma_{12} & \sigma_{13} & \cdots & \sigma_{1p} \\
\sigma_{12} & \sigma_2^2 & \sigma_{23} & \cdots & \sigma_{2p} \\
\sigma_{13} & \sigma_{23} & \sigma_3^2 & \cdots & \sigma_{3p} \\
\cdots & \cdots & \cdots & \cdots & \cdots \\
\sigma_{1p} & \sigma_{2p} & \sigma_{3p} & \cdots & \sigma_p^2
\end{bmatrix}.
$$

This matrix has a symmetric structure with p variances and $p(p-1)/2$ covariances, for a total of $p(p+1)/2$ parameters that have to be estimated from the data. If the size of the data set permits, the model with unstructured variance-covariance matrix of the error terms should be fitted. This model doesn't make any assumptions regarding the structure and allows the estimate of each parameter to take on its own value.

If the size of the data set is too small to fit a general unstructured variance-covariance matrix, there are several sparse but meaningful structures that may be tried. Here we present four most commonly used models.

• The first model has the blocks in the variance-covariance matrix that have constant values on each of the descending diagonals, that is, the matrix has the form:

$$\begin{bmatrix} \sigma^2 & \sigma_1 & \sigma_2 & \cdots & \sigma_{p-1} \\ \sigma_1 & \sigma^2 & \sigma_1 & \cdots & \sigma_{p-2} \\ \sigma_2 & \sigma_1 & \sigma^2 & \cdots & \sigma_{p-3} \\ \cdots & \cdots & \cdots & \cdots & \cdots \\ \sigma_{p-1} & \sigma_{p-2} & \sigma_{p-3} & \cdots & \sigma^2 \end{bmatrix}.$$

Using this structure implies that observations that are the same number of time points apart are equally correlated. There are a total of p unknown parameters $\sigma^2, \sigma_1, \ldots, \sigma_{p-1}$. This model is said to have the *Toeplitz* covariance structure, which is sometimes referred to as an autoregressive-moving average ARMA(1,1) structure.

• Another model with useful structure of the variance-covariance matrix relies on the fact that typically as time goes on, observations become less correlated with the earlier ones. In this model, each block in the variance-covariance matrix has $\sigma^2 \rho^{|t_i - t_j|}$ in the ij-th cell, $i, j = 1, \ldots, p$, that is, it looks like this:

$$\sigma^2 \begin{bmatrix} 1 & \rho^{|t_1 - t_2|} & \rho^{|t_1 - t_3|} & \cdots & \rho^{|t_1 - t_p|} \\ \rho^{|t_1 - t_2|} & 1 & \rho^{|t_2 - t_3|} & \cdots & \rho^{|t_2 - t_p|} \\ \rho^{|t_1 - t_3|} & \rho^{|t_2 - t_3|} & 1 & \cdots & \rho^{|t_3 - t_p|} \\ \cdots & \cdots & \cdots & \cdots & \cdots \\ \rho^{|t_1 - t_p|} & \rho^{|t_2 - t_p|} & \rho^{|t_3 - t_p|} & \cdots & 1 \end{bmatrix}.$$

Here σ^2 and ρ are the unknown constants, $|\rho| < 1$. Note that in this matrix the entries decrease as the distance between times t_i and t_j increases. The matrix of this form is termed a *spatial power matrix*, and hence the model is said to have a *spatial power* variance-covariance structure of the error terms.

• A special case of this model is when the times are equal to $1, 2, 3, \ldots, p$. Then the $p \times p$ blocks of the variance-covariance matrix become:

$$\sigma^2 \begin{bmatrix} 1 & \rho & \rho^2 & \cdots & \rho^{p-1} \\ \rho & 1 & \rho & \cdots & \rho^{p-2} \\ \rho^2 & \rho & 1 & \cdots & \rho^{p-3} \\ \cdots & \cdots & \cdots & \cdots & \cdots \\ \rho^{p-1} & \rho^{p-2} & \rho^{p-3} & \cdots & 1 \end{bmatrix}.$$

This model is said to have an *autoregressive* variance-covariance structure of the error terms, referring to an AR(1) model, an autoregressive time series model with lag one that has the same covariance structure. Note that the autoregressive matrix is a special case of both Toeplitz and spatial power matrices.

• The last model that we introduce here is the model with *compound symmetric* (or *exchangeable*) variance-covariance matrix of the error terms. In this matrix, all variances are assumed to be equal to σ^2, and all correlations are assumed to be equal to ρ. That is, the matrix has the form:

$$\sigma^2 \begin{bmatrix} 1 & \rho & \rho & \cdots & \rho \\ \rho & 1 & \rho & \cdots & \rho \\ \rho & \rho & 1 & \cdots & \rho \\ \cdots & \cdots & \cdots & \cdots & \cdots \\ \rho & \rho & \rho & \cdots & 1 \end{bmatrix}.$$

This variance-covariance structure better suits situations when repeated measurements are taken under various treatment conditions rather than longitudinally since all observations are assumed equally correlated.

8.2.2 Fitted Model, Interpretation of Estimated Regression Coefficients, and Predicted Response

The fitted model has the same form as the fitted slope and intercept model with independent covariance structure for errors (8.3). Estimated regression coefficients are interpreted the same way, and predicted responses for fixed sets of predictors are computed similarly (see Subsections 8.1.2, 8.1.3, and 8.1.5).

8.2.3 Model Goodness-of-fit Check

Goodness-of-fit of several models are compared based on AIC, AICC, and BIC criteria.

For a given model, the goodness of fit may be verified by carrying out the chi-squared deviance test. In this case, there are no random-effects terms in the null model, thus it has $k + 3$ parameters: $\beta_0, \ldots, \beta_{k+1}$, and σ. The fitted model has $k+2$ beta parameters, 3 random-effects parameters $\sigma^2_{u_1}$, $\sigma_{u_1 u_2}$, and $\sigma^2_{u_2}$, plus the number of parameters in the variance-covariance matrix of error terms. Since the number of degrees of freedom is calculated as the difference between the number of parameters in the fitted and null models, the number of degrees of freedom is: for unstructured covariance matrix, $df = k + 5 + \dfrac{p(p+1)}{2} - (k+3) =$

$\dfrac{p(p+1)}{2} + 2$, for Toeplitz matrix, $df = k + 5 + p - (k+3) = p + 2$, and for spatial power, autoregressive, and compound symmetric matrices, $df = k + 5 + 2 - (k+3) = 4$.

8.2.4   SAS Implementation

The random slope and intercept model with a certain covariance structure for the error terms is fitted in SAS by running `proc mixed` described in Subsection 8.1.6 with the `repeated` statement added as follows:

`repeated / subject=`*id_name* `type=`*covtype_name* `r;`

• Covariance structures in the option `type=` are `un` (unstructured), `toep` (Toeplitz), `sp(pow)`(*time_name*) (spatial power), `ar(1)` (autoregressive), or `cs` (compound symmetric).
• By specifying the option `r` we request that SAS outputs an estimated individual block of the variance-covariance matrix of the error terms, labeling it `Estimated R Matrix for Subject 1`.
• Estimated correlation coefficient ρ in the output is labeled `SP(POW)` (spatial power), `AR(1)` (autoregressive), or `CS` (compound symmetric).

8.2.5   R Implementation

The syntax for the random slope and intercept model with the independent structure for errors (see Subsection 8.1.7) is valid for all the other structures.
• A structure can be specified by adding `correlation=` to the syntax within the `lme()` function. The choices are:

`corSymm()` (unstructured),
`corARMA(form=`~ `1|`*id.name*`, p=1, q=1)` (Toeplitz),
`corCAR1(form=`~`1|`*id.name*`)` (spatial power),
`corAR1(form=`~`1|`*id.name*`)` (autoregressive),
and
`corCompSymm(form=`~`1|`*id.name*`)` (compound symmetric).

• In the model with an unstructured variance-covariance matrix, by default, R produces identical estimates of the variances. To lift this restriction, one should add to the function `lme()` the argument `weights=varIdent(form`~ *id.name*|*time.name*).
• A fitted individual block of the variance-covariance matrix can be viewed by

typing on a separate line getVarCov(*fitted.model.name*, type="conditional").

8.2.6   Example

EXAMPLE 8.2. In Example 8.1, we fitted random slope and intercept model with the independent structure of the error terms. The outputted goodness-of-fit statistics were AIC=870.3, AICC=870.7, and BIC=875.5. Below we run models with the unstructured, Toeplitz, spatial power, autoregressive, and compound symmetric structures, and compare their fit statistics.

```
/*fitting random slope and intercept model with
unstructured covariance matrix of error terms*/
proc mixed covtest;
 class gender;
  model LDL=gender age month/solution;
   random intercept month/subject=id type=un;
 repeated/subject=id type=un r;
run;
```

In this model, all estimated matrix entries are all different:

Estimated R Matrix for Subject 1

Row	Col1	Col2	Col3	Col4
1	398.30	339.18	189.07	101.14
2	339.18	355.61	190.89	90.9265
3	189.07	190.89	222.62	141.82
4	101.14	90.9265	141.82	320.07

The variances and covariance of the random terms are given below. Note that they are not statistically significant at the 5% level.

Covariance Parameter Estimates

Cov Parm	Estimate	Pr Z
UN(1,1)	185.94	0.1593
UN(2,1)	-5.2666	0.4776
UN(2,2)	0.02170	0.4757

The goodness-of-fit criteria values and the results for the asymptotic likelihood ratio test are:

Fit Statistics
AIC   874.7
AICC   878.7
BIC   891.5

Null Model Likelihood Ratio Test
DF Chi-Square   Pr > ChiSq
12   86.49   <.0001

The estimated coefficients for the fixed-effects terms are:

Solution for Fixed Effects

Effect	gender	Estimate	Pr > |t|
Intercept		60.6156	0.0039
gender	F	28.5591	<.0001
gender	M	0	.
age		0.9601	0.0049
month		-1.0250	<.0001

```
/*fitting random slope and intercept model with
Toeplitz covariance matrix of error terms*/
proc mixed covtest;
 class gender;
  model LDL=gender age month/solution;
   random intercept month/subject=id type=un;
 repeated / subject=id type=toep r;
run;
```

For each individual, the fitted variance-covariance block is:

Estimated R Matrix for Subject 1

Row	Col1	Col2	Col3	Col4
1	89.3298	26.1821	-11.1480	38.7476
2	26.1821	89.3298	26.1821	-11.1480
3	-11.1480	26.1821	89.3298	26.1821
4	38.7476	-11.1480	26.1821	89.3298

Note that in this matrix the values on each descending diagonal are the same.

Further, the parameters for the random terms are statistically significant since the respective p-values are less than 0.05, as seen in the following output:

Covariance Parameter Estimates

Cov Parm	Estimate	Pr Z
UN(1,1)	466.45	0.0016
UN(2,1)	-13.9159	0.0206
UN(2,2)	0.6376	0.0239

The model fit results are given as:

Fit Statistics

AIC	871.2
AICC	872.3
BIC	880.2

Null Model Likelihood Ratio Test

DF	Chi-Square	Pr > ChiSq
6	78.03	<.0001

The estimated beta coefficients for the fixed-effects terms are:

Solution for Fixed Effects

Effect	gender	Estimate	Pr > |t|
Intercept		61.1870	0.0041
gender	F	29.5657	<.0001
gender	M	0	.
age		0.9662	0.0053
month		-0.9946	<.0001

```
/*fitting random slope and intercept model with
spatial power covariance matrix of error terms*/
proc mixed covtest;
  class gender;
   model LDL=gender age month/solution;
    random intercept month/subject=id type=un;
  repeated / subject=id type=sp(pow)(month) r;
run;
```

Each estimated block in the variance-covariance matrix is:

Estimated R Matrix for Subject 1

Row	Col1	Col2	Col3	Col4
1	69.8556	3.14E-12	0	0
2	3.14E-12	69.8556	0.000015	0
3	0	0.000015	69.8556	0
4	0	0	0	69.8556

In this model, the correlation coefficient ρ is statistically indistinguishable from zero. The times of measurements are $t_1 = 0$, $t_2 = 6$, $t_3 = 9$, and $t_4 = 24$. The differences between these times assume values 3, 6, 9, 15, 18, and 24, and thus the off-diagonal entries in the matrix are: $\hat{\sigma}^2 \hat{\rho}^3$, $\hat{\sigma}^2 \hat{\rho}^6$, $\hat{\sigma}^2 \hat{\rho}^9$, $\hat{\sigma}^2 \hat{\rho}^{15}$, $\hat{\sigma}^2 \hat{\rho}^{18}$, $\hat{\sigma}^2 \hat{\rho}^{24}$, which are all estimated to be very close to zero.

The random-effects parameters are estimated as:

Covariance Parameter Estimates

Cov Parm	Estimate	Pr Z
UN(1,1)	520.17	0.0013
UN(2,1)	-16.3953	0.0087
UN(2,2)	0.7846	0.0028
SP(POW)	0.005962	0.9985
Residual	69.8556	<.0001

The summary of the model fit is given below.

Fit Statistics

AIC	872.3
AICC	872.9
BIC	878.8

Null Model Likelihood Ratio Test

DF	Chi-Square	Pr > ChiSq
4	72.86	<.0001

The estimated regression coefficients for the fixed-effects terms are:

Solution for Fixed Effects

Effect	gender	Estimate	Pr > |t|
Intercept		65.0163	0.0029
gender	F	29.8105	<.0001
gender	M	0	.
age		0.9203	0.0085
month		-1.0957	<.0001

```
/*fitting random slope and intercept model with
autoregressive covariance matrix of error terms*/
proc mixed covtest;
  class gender;
    model LDL=gender age month/solution;
```

227

```
   random intercept month/subject=id type=un;
 repeated / subject=id type=ar(1) r;
run;
```

Each estimated block in the variance-covariance matrix is:

```
    Estimated R Matrix for Subject 1
Row   Col1     Col2      Col3      Col4
1    157.02   87.6923   48.9756   27.3526
2    87.6923  157.02    87.6923   48.9756
3    48.9756  87.6923   157.02    87.6923
4    27.3526  48.9756   87.6923   157.02
```

The estimates $\widehat{\sigma}^2 = 157.02$ and $\widehat{\rho} = 0.5585$. The entries in the estimated matrix are $\widehat{\sigma}^2 = 157.02$, $\widehat{\sigma}^2\widehat{\rho} = (157.02)(0.5585) = 87.69$, $\widehat{\sigma}^2\widehat{\rho}^2 = (157.02)(0.5585^2) = 48.97$, and $\widehat{\sigma}^2\widehat{\rho}^3 = (157.02)(0.5585^3) = 27.35$.

The random-effects parameters are estimated as:

```
    Covariance Parameter Estimates
Cov Parm   Estimate     Pr Z
UN(1,1)     388.12      0.0365
UN(2,1)    -11.7246     0.0610
UN(2,2)     0.4812      0.0527
AR(1)       0.5585      0.0816
Residual    157.02      0.1019
```

The model fit statistics are given below.

```
Fit Statistics
AIC    869.6
AICC   870.2
BIC    876.0
```

```
Null Model Likelihood Ratio Test
DF Chi-Square  Pr > ChiSq
 4     75.63       <.0001
```

The estimated beta coefficients for the fixed-effects terms are:

Solution for Fixed Effects

Effect	gender	Estimate	Pr > |t|
Intercept		62.9465	0.0044
gender	F	30.6095	<.0001
gender	M	0	.
age		0.9378	0.0088
month		-0.9640	<.0001

```
/*fitting random slope and intercept model with
compound symmetric covariance matrix of error terms*/
proc mixed covtest;
 class gender;
  model LDL=gender age month/solution;
   random intercept month/subject=id type=un;
 repeated / subject=id type=cs r;
run;
```

For each individual, the fitted block is:

Estimated R Matrix for Subject 1

Row	Col1	Col2	Col3	Col4
1	102.03	32.1771	32.1771	32.1771
2	32.1771	102.03	32.1771	32.1771
3	32.1771	32.1771	102.03	32.1771
4	32.1771	32.1771	32.1771	102.03

Note that this matrix has identical off-diagonal elements.

The estimates of the variance-covariance parameters for the random terms are as follows:

Covariance Parameter Estimates

Cov Parm	Estimate	Pr Z
UN(1,1)	487.99	0.0024
UN(2,1)	-16.3952	0.0087
UN(2,2)	0.7846	0.0028
CS	32.1771	<.0001
Residual	69.8558	<.0001

The goodness-of-fit related output is:

Fit Statistics
AIC   872.3
AICC   872.9
BIC   878.8

Null Model Likelihood Ratio Test
DF   Chi-Square   Pr > ChiSq
 4   72.86   <.0001

The estimated regression coefficients for the fixed-effects terms are:

Solution for Fixed Effects

Effect	gender	Estimate	Pr > |t|
Intercept		65.0161	0.0029
gender	F	29.8106	<.0001
gender	M	0	.
age		0.9203	0.0085
month		-1.0957	<.0001

To identify the best-fitted model, we compare the values for AIC, AICC, and BIC criteria and identify the smallest ones. We summarize them here:

	Ind	Un	Toep	SpPow	AR	Exch
AIC	870.3	874.7	871.2	872.3	869.6	872.3
AICC	870.7	878.7	872.3	872.9	870.2	872.9
BIC	875.5	891.5	880.2	878.8	876.0	878.8

According to AIC and AICC criteria, the model with autoregressive structure is optimal, whereas the BIC criterion dictates that the model with the independent structure is better.

The fitted model with the autoregressive structure is $\widehat{E}(LDL) = 62.9465 + 30.6095 * female + 0.9378 * age - 0.9640 * month$. All the fixed-effects predictors are significant at the 0.05 level, whereas only the variance of the random slope is significant at that level. On average, the estimated cholesterol level is 30.6095 points higher for female than for male patients. As age increases by one year, the estimated mean LDL increases by 0.9203 points, and it decreases by 1.0957 points for every additional month in the study.

To compute the predicted LDL level for a 48-year-old female patient three months into the study, we write $LDL^0 = 62.9465 + 30.6095 + 0.9378 * 48 - 0.9640 * 3 = 135.6784$. SAS outputs the same result:

```
data prediction;
input id gender$ age month;
cards;
28 F 48 3
;

data longform;
 set longform prediction;
run;

proc mixed covtest;
 class gender;
  model LDL=gender age month/solution outpm=outdata;
   random intercept month/subject=id type=un;
 repeated/subject=id type=ar(1) r;
run;

proc print data=outdata (firstobs=109 obs=109);
var Pred;
run;
```

```
    Pred
135.678
```

The R codes and outputs for the long-form data set in this example follow.

```
install.packages("nlme")
library(nlme)

#fitting random slope and intercept model with
#unstructured covariance matrix of error terms
summary(un.fitted.model<-lme(LDL ~ gender.rel + age + month,
random =~ 1 + month|id,data=longform.data,
correlation=corSymm(), weights=varIdent(form=~ id|month)))
getVarCov(un.fitted.model, type="conditional")
```

```
   AIC       BIC      logLik
882.703   927.6576   -424.3515
Random effects:
             StdDev      Corr
(Intercept) 22.1065205   (Intr)
month        0.8202094   -0.767
```

Fixed effects:

	Value	p-value
(Intercept)	60.61574	0.0020
gender.relF	28.55931	0.0003
age	0.96009	0.0072
month	-1.02503	0.0000

Conditional variance covariance matrix

	1	2	3	4
1	95.5290	57.655	-35.968	8.6119
2	57.6550	56.814	-11.431	-46.5200
3	-35.9680	-11.431	22.571	-25.3850
4	8.6119	-46.520	-25.385	133.0400

```
# computing AICC
n<- 108
p<- 17
print(AICC<- -2*logLik(un.fitted.model)+2*p*n/(n-p-1))
```

889.503

```
#checking model fit
summary(null.model<- glm(LDL ~ gender.rel + age + month,
data=longform.data, family=gaussian(link=identity)))
print(deviance<- -2*(logLik(null.model)
  -logLik(un.fitted.model)))
```

91.43229

```
print(p.value<- pchisq(deviance, df=12, lower.tail=FALSE))
```

2.605906e-14

```
#fitting random slope and intercept model with
#Toeplitz covariance matrix of error terms
summary(toep.fitted.model<- lme(LDL ~ gender.rel + age + month,
random=~ 1 + month|id, data=longform.data,
correlation=corARMA(form=~ 1|id, p=1, q=1)))
getVarCov(toep.fitted.model, type="conditional")
```

AIC	BIC	logLik
877.5213	903.9653	-428.7607

Random effects:

	StdDev	Corr

```
(Intercept)  21.0541831     (Intr)
month         0.7154187    -0.813
Fixed effects:
              Value      p-value
(Intercept)  61.41936    0.0024
gender.relF  30.11579    0.0002
age           0.95896    0.0091
month        -0.95868    0.0000
Conditional variance covariance matrix
           1        2        3        4
1 101.4400  35.200  -10.762   3.2906
2  35.2000 101.440   35.200 -10.7620
3 -10.7620  35.200  101.440  35.2000
4   3.2906 -10.762   35.200 101.4400
```

```
#computing AICC
p<- 11
print(AICC<- -2*logLik(toep.fitted.model)+2*p*n/(n-p-1))
```

882.2713

```
#checking model fit
print(deviance<- -2*(logLik(null.model)
-logLik(toep.fitted.model)))
```

82.61391

```
print(p.value<- pchisq(deviance, df=6, lower.tail=FALSE))
```

1.029563e-15

```
#fitting random slope and intercept model with
#spatial power covariance matrix of error terms
summary(sppow.fitted.model<- lme(LDL ~ gender.rel + age + month,
random=~ 1 + month|id, data=longform.data,
correlation=corCAR1(form=~ month|id)))
getVarCov(sppow.fitted.model, type="conditional")
```

```
    AIC        BIC       logLik
880.3294    904.1289    -431.1647
Random effects:
               StdDev        Corr
(Intercept)  22.8068074     (Intr)
```

```
month         0.8857947     -0.812
Residual      8.3578134
```

Parameter estimate:

```
       Phi
0.005521199
```

Fixed effects:

```
                Value      p-value
(Intercept) 65.01447       0.0014
gender.relF 29.81150       0.0003
age          0.92035       0.0116
month       -1.09566       0.0000
```

Conditional variance covariance matrix

	1	2	3	4
1	6.9853e+01	1.9787e-12	3.3303e-19	4.4977e-53
2	1.9787e-12	6.9853e+01	1.1757e-05	1.5878e-39
3	3.3303e-19	1.1757e-05	6.9853e+01	9.4338e-33
4	4.4977e-53	1.5878e-39	9.4338e-33	6.9853e+01

```
#computing AICC
p<- 9
print(AICC<- -2*logLik(sppow.fitted.model)+2*p*n/(n-p-1))
```

882.1661

```
#checking model fit
print(deviance<- -2*(logLik(null.model)
-logLik(sppow.fitted.model)))
```

77.80585

```
print(p.value<- pchisq(deviance, df=4, lower.tail=FALSE))
```

5.077837e-16

```
#fitting random slope and intercept model with
#autoregressive covariance matrix of error terms
summary(ar1.fitted.model<- lme(LDL ~ gender.rel + age + month,
random=~ 1 + month|id, data=longform.data,
correlation=corAR1(form=~ 1|id)))
getVarCov(ar1.fitted.model, type="conditional")
```

```
    AIC       BIC      logLik
877.5555   901.355   -429.7777
```

Random effects:
```
              StdDev         Corr
(Intercept)  19.7010538     (Intr)
month         0.6936938    -0.858
Residual     12.5304689
```
Parameter estimate:
```
      Phi
0.5584911
```
Fixed effects:
```
                Value     p-value
(Intercept) 62.94677      0.0023
gender.relF 30.60932      0.0003
age          0.93779      0.0120
month       -0.96402      0.0000
```
Conditional variance covariance matrix
```
          1        2        3        4
1  157.010   87.690   48.974   27.352
2   87.690  157.010   87.690   48.974
3   48.974   87.690  157.010   87.690
4   27.352   48.974   87.690  157.010
```

```
#computing AICC
p<- 9
print(AICC<- -2*logLik(ar1.fitted.model)+2*p*n/(n-p-1))
```

879.3922

```
#checking model fit
print(deviance<- -2*(logLik(null.model)
-logLik(ar1.fitted.model)))
```

80.57979

```
print(p.value<- pchisq(deviance, df=4, lower.tail=FALSE))
```

1.312697e-16

```
#fitting random slope and intercept model with
#compound symmetry covariance matrix of error terms
summary(cs.fitted.model<- lme(LDL ~ gender.rel + age + month,
random=~ 1 + month|id, data=longform.data,
correlation=corCompSymm(form=~ 1|id)))
getVarCov(cs.fitted.model, type="conditional")
```

```
       AIC        BIC       logLik
   880.3294   904.1289   -431.1647
Parameter estimate:
Rho
  0
Random effects:
                StdDev      Corr
(Intercept)  22.8072879    (Intr)
month         0.8857844   -0.812
Residual      8.3579458
Fixed effects:
                Value    p-value
(Intercept)  65.01614     0.0014
gender.relF  29.81056     0.0003
age           0.92033     0.0116
month        -1.09566     0.0000
Conditional variance covariance matrix
          1       2       3       4
1  69.855   0.000   0.000   0.000
2   0.000  69.855   0.000   0.000
3   0.000   0.000  69.855   0.000
4   0.000   0.000   0.000  69.855
```

```
#computing AICC
p<- 9
print(AICC<- -2*logLik(cs.fitted.model)+2*p*n/(n-p-1))
```

882.1661

```
#checking model fit
print(deviance<- -2*(logLik(null.model)
-logLik(cs.fitted.model)))
```

77.80585

```
print(p.value<- pchisq(deviance, df=3, lower.tail=FALSE))
```

5.077834e-16

Next, we summarize the fit statistics to decide which model is optimal.

	Ind	Un	Toep	SpPow	AR	Exch
AIC	878.3	882.7	877.5	880.3	877.6	880.3
AICC	879.8	889.5	882.3	882.2	879.4	882.2
BIC	899.5	927.7	904.0	904.1	901.4	904.1

Here we make the same conclusion as in SAS: AIC and AICC criteria choose the autoregressive structure, while the independent structure is better with respect to the BIC criterion.

Finally, similar to the analysis done in SAS, we use the model with autoregressive covariance structure for prediction. The script and output follow.

```
#using AR fitted model for prediction
print(predict(ar1.fitted.model, data.frame(gender.rel="F",
age=48, month=3), level=0))
```

135.6782

□

8.3 Generalized Estimating Equations Model

8.3.1 Model Definition

So far in this chapter we considered modeling data with repeated measures by a random slope and intercept model with a specified structure of the variance-covariance matrix of the error terms. Another way to model repeated measures data is to use a *Generalized Estimating Equations*[2] (GEE) approach. In the GEE model, there are no random-effects terms. Instead, the response variable relates to the predictors via the generalized linear regression model with only fixed-effects terms, and the variance-covariance structure of the response variable itself is specified, rather than that for the error terms.

Suppose that for each individual i, $i = 1, \ldots, n$, at time t_j, $j = 1, \ldots, p$, the observations are $x_{1ij}, \ldots, x_{kij}, y_{ij}$. Denote by $\mu_{ij} = \mathbb{E}(y_{ij})$, the *mean response*, and assume that $g(\mu_{ij}) = \beta_0 + \beta_1 x_{1ij} + \cdots + \beta_k x_{kij} + \beta_{k+1} t_j$ where $g(\cdot)$ is the link function.

Next, we write the variance of y_{ij} as a function of μ_{ij}, $\mathbb{V}ar(y_{ij}) = V(\mu_{ij})$. The function $V(\cdot)$ is termed the *variance function*.

Further, we model the covariance structure of correlated responses for a given individual i, $i = 1, \ldots, n$. Observations between individuals are assumed independent. Let $\mathbf{A}_i$ denote a $p \times p$ diagonal matrix

---

[2]Originally discussed in Liang, K.-Y. and S.L. Zeger (1986). "Longitudinal Data Analysis Using Generalized Linear Models". *Biometrika*, 73(1): 13 – 22, and in Zeger, S.L. and K.-Y. Liang (1986). "Longitudinal Data Analysis for Discrete and Continuous Outcomes". *Biometrics*, 42(1): 121 – 130.

$$\mathbf{A}_i = \begin{bmatrix} V(\mu_{i1}) & 0 & \cdots & 0 \\ 0 & V(\mu_{i2}) & \cdots & 0 \\ \cdots & \cdots & \cdots & \cdots \\ 0 & 0 & \cdots & V(\mu_{ip}) \end{bmatrix}.$$

Also, let $\mathbf{R}_i(\boldsymbol{\alpha})$ be the *working correlation matrix* of the repeated responses for the i-th individual, where $\boldsymbol{\alpha}$ denotes a vector of unknown parameters which are the same for all individuals. The working covariance matrix for $\mathbf{y}_i = (y_{i1}, y_{i2}, \ldots, y_{ip})'$ is then

$$\mathbf{V}_i(\boldsymbol{\alpha}) = \mathbf{A}_i^{1/2} \mathbf{R}_i(\boldsymbol{\alpha}) \mathbf{A}_i^{1/2}.$$

The regression coefficients β's and the vector of parameters $\boldsymbol{\alpha}$ are the only unknowns of the GEE model and must be estimated from the data. Five structures are commonly used for the working correlation matrix $\mathbf{R}_i(\boldsymbol{\alpha})$. They are:

- unstructured (with all different off-diagonal entries, a total of $p(p-1)/2$ unknown parameters)

$$\mathbf{R}_i(\boldsymbol{\alpha}) = \begin{bmatrix} 1 & \alpha_{12} & \alpha_{13} & \cdots & \alpha_{1p} \\ \alpha_{12} & 1 & \alpha_{23} & \cdots & \alpha_{2p} \\ \alpha_{13} & \alpha_{23} & 1 & \cdots & \alpha_{3p} \\ \cdots & \cdots & \cdots & \cdots & \cdots \\ \alpha_{1p} & \alpha_{2p} & \alpha_{3p} & \cdots & 1 \end{bmatrix},$$

- Toeplitz (with identical entries on each descending diagonal, a total of $p-1$ unknown parameters)

$$\mathbf{R}_i(\boldsymbol{\alpha}) = \begin{bmatrix} 1 & \alpha_1 & \alpha_2 & \cdots & \alpha_{p-1} \\ \alpha_1 & 1 & \alpha_1 & \cdots & \alpha_{p-2} \\ \alpha_2 & \alpha_1 & 1 & \cdots & \alpha_{p-3} \\ \cdots & \cdots & \cdots & \cdots & \cdots \\ \alpha_{p-1} & \alpha_{p-2} & \alpha_{p-3} & \cdots & 1 \end{bmatrix},$$

- autoregressive (with $\alpha^{|i-j|}$ in the ij-th position, a total of one unknown parameter)

$$\mathbf{R}_i(\boldsymbol{\alpha}) = \begin{bmatrix} 1 & \alpha & \alpha^2 & \cdots & \alpha^{p-1} \\ \alpha & 1 & \alpha & \cdots & \alpha^{p-2} \\ \alpha^2 & \alpha & 1 & \cdots & \alpha^{p-3} \\ \cdots & \cdots & \cdots & \cdots & \cdots \\ \alpha^{p-1} & \alpha^{p-2} & \alpha^{p-3} & \cdots & 1 \end{bmatrix},$$

- compound symmetric or exchangeable (with all identical off-diagonal elements, a total of one unknown parameter)

$$\mathbf{R}_i(\boldsymbol{\alpha}) = \begin{bmatrix} 1 & \alpha & \alpha & \dots & \alpha \\ \alpha & 1 & \alpha & \dots & \alpha \\ \alpha & \alpha & 1 & \dots & \alpha \\ \dots & \dots & \dots & \dots & \dots \\ \alpha & \alpha & \alpha & \alpha & 1 \end{bmatrix},$$

- independent (identity matrix, no unknown parameters)

$$\mathbf{R}_i(\boldsymbol{\alpha}) = \begin{bmatrix} 1 & 0 & 0 & \dots & 0 \\ 0 & 1 & 0 & \dots & 0 \\ 0 & 0 & 1 & \dots & 0 \\ \dots & \dots & \dots & \dots & \dots \\ 0 & 0 & 0 & \dots & 1 \end{bmatrix}.$$

The GEE estimate of $\boldsymbol{\beta} = (\beta_0, \beta_1, \dots, \beta_{k+1})'$ is the solution of the *generalized estimating equations*:

$$\sum_{i=1}^{n} \left(\frac{\partial \boldsymbol{\mu}_i}{\partial \boldsymbol{\beta}} \right)_{(k+2) \times p} \left[\mathbf{V}_i(\hat{\boldsymbol{\alpha}}) \right]^{-1}_{p \times p} \left(\mathbf{y}_i - \boldsymbol{\mu}_i \right)_{p \times 1} = \mathbf{0}_{(k+2) \times 1}$$

where $\boldsymbol{\mu}_i = (\mu_{i1}, \dots, \mu_{ip})'$ is the vector of mean responses, and the estimator $\hat{\boldsymbol{\alpha}}$ is the method of moments estimator of the vector of parameters.

8.3.2 Fitted Model, Interpretation of Estimated Regression Coefficients, and Predicted Response

In this chapter we consider the GEE model with a normally distributed response, hence, the link function is the identity function, $\mathbb{E}(y_{ij}) = \beta_0 + \beta_1 x_{1ij} + \dots + \beta_k x_{kij} + \beta_{k+1} t_j$. The fitted model, interpretation of beta estimates, and prediction are the same as in the random slope and intercept model (see Subsections 8.1.2, 8.1.3, and 8.1.5).

8.3.3 Model Goodness-of-Fit Check

The *quasi-likelihood under the independence* (QIC) model criterion statistic[3] is used as the goodness-of-fit measure. It is based on the *quasi-likelihood function* Q defined as follows:

$$Q = \sum_{i=1}^{n} \sum_{j=1}^{p} \int_{y_{ij}}^{\mu_{ij}} \frac{y_{ij} - u}{V(u)} \, du.$$

The QIC is used to select the best-fitted working correlation structure. The one with the smallest value wins.

---

[3]Introduced in Pan, W. (2001). "Akaike's Information Criterion in Generalized Estimating Equations". *Biometrics*, 57(1): 120 – 125.

8.3.4 SAS Implementation

To fit a GEE model, one can use the `genmod` procedure with the `repeated` statement. The syntax is below.

```
proc genmod data=longform_data_name;
  class <list of categorical predictors>;
    model response_name=<list of x predictors> time_name/
                          dist=normal link=identity;
      output out=outdata_name p=predicted_response_name;
      repeated subject=id_name/type=corrtype_name corrw covb;
run;
```

- The option `dist=normal link=identity` is a default in this procedure and may be omitted.
- The types of the working correlation matrix are `un` for unstructured, `mdep(m)` for Toeplitz with m repeated observations for each individual, `ar` for autoregressive, `cs` or `exch` for compound symmetric (or exchangeable), and `ind` for independent.
- Specifying the option `corrw` produces the estimated working correlation matrix for a single individual.

8.3.5 R Implementation

Function `geeglm()` in package `geepack` may be employed to fit a GEE model with the specified underlying distribution. In this chapter we model only the normally distributed response. Other distributions will be considered in the next chapter. This function should be run on a long-form data set that is sorted by individuals' id. The syntax for sorting is:

longform.data.name<- *longform.data.name*
[order(*longform.data.name$id.name*),]

The syntax for fitting GEE models is as follows:

`summary(`*fitted.model.name*<- `geeglm(`*response.name* ~ *x1.name* + ⋯
+ *xk.name* + *time.name*, `data=`*longform.data.name*, `id=`*id.name*,
`family=gaussian(link="identity")`, `corstr="`*corr.structure.name*`"))`

- R doesn't automatically fit models with the Toeplitz correlation matrix. The choices for the correlation structures are `unstructured`, `ar1`, `exchangeable`, and `independence`.

• To request the value of QIC, one should type `QIC(`*fitted.model.name*`)`, preliminary installing the package `MuMIn`, which performs multi-model inference.

8.3.6 Example

EXAMPLE 8.3. We will use the data in Example 8.1 to fit GEE models with various working correlation matrices, and select the best-fitted one. The SAS code and relevant outputs follow.

```
/* fitting GEE model with unstructured working correlation
matrix */
proc genmod;
class id gender;
 model ldl=gender age month/dist=normal link=identity;
  repeated subject=id/type=un corrw;
run;
```

    WARNING: Iteration limit exceeded.

    Working Correlation Matrix
          Col1   Col2   Col3   Col4
Row1 1.0000 0.9810 0.9810 0.8160
Row2 0.9810 1.0000 0.9810 0.7583
Row3 0.9810 0.9810 1.0000 0.6895
Row4 0.8160 0.7583 0.6895 1.0000

        GEE Fit Criteria
        QIC 144.1917

Analysis Of GEE Parameter Estimates
Parameter   Estimate   Pr > |Z|
Intercept   403.9145   0.0763
gender   F   -163.042   0.0354
gender   M   0.0000   .
age   -2.7167   0.4942
month   -1.5905   <.0001

Note that the algorithm doesn't converge, and therefore we eliminate this model from the list of candidates.

```
/* fitting GEE model with Toeplitz working correlation matrix */
proc genmod;
class id gender;
 model ldl=gender age month/dist=normal link=identity;
repeated subject=id /type=mdep(3) corrw;
run;
```

Working Correlation Matrix
	Col1	Col2	Col3	Col4
Row1	1.0000	0.7536	0.4798	0.3084
Row2	0.7536	1.0000	0.7536	0.4798
Row3	0.4798	0.7536	1.0000	0.7536
Row4	0.3084	0.4798	0.7536	1.0000

GEE Fit Criteria
QIC 119.1042

Analysis Of GEE Parameter Estimates
Parameter		Estimate	Pr > $|Z|$
Intercept		54.1516	0.0028
gender	F	36.0994	<.0001
gender	M	0.0000	.
age		1.0174	0.0010
month		-0.8347	<.0001

```
/* fitting GEE model with autoregressive working
correlation matrix*/
proc genmod;
class id gender;
 model ldl=gender age month/dist=normal link=identity;
  repeated subject=id /type=ar corrw covb;
run;
```

Working Correlation Matrix
	Col1	Col2	Col3	Col4
Row1	1.0000	0.7523	0.5659	0.4257
Row2	0.7523	1.0000	0.7523	0.5659
Row3	0.5659	0.7523	1.0000	0.7523
Row4	0.4257	0.5659	0.7523	1.0000

GEE Fit Criteria
QIC 119.4599

Analysis Of GEE Parameter Estimates
Parameter   Estimate   Pr > |Z|
Intercept   54.0026   0.0036
gender   F   36.3525   <.0001
gender   M   0.0000   .
age   1.0274   0.0012
month   -0.9119   <.0001

```
/*fitting GEE with compound symmetric working
correlation matrix*/
proc genmod;
class id gender;
model ldl=gender age month/dist=normal link=identity;
repeated subject=id /type=cs corrw;
run;
```

Working Correlation Matrix
        Col1   Col2   Col3   Col4
Row1 1.0000 0.5743 0.5743 0.5743
Row2 0.5743 1.0000 0.5743 0.5743
Row3 0.5743 0.5743 1.0000 0.5743
Row4 0.5743 0.5743 0.5743 1.0000

GEE Fit Criteria
QIC 120.8579

Analysis Of GEE Parameter Estimates
Parameter   Estimate   Pr > |Z|
Intercept   51.5122   0.0122
gender   F   37.4045   <.0001
gender   M   0.0000   .
age   1.0692   0.0024
month   -1.0957   <.0001

```
/*GEE with independent working correlation matrix*/
proc genmod;
class id gender;
model ldl=gender age month/dist=normal link=identity;
repeated subject=id /type=ind corrw;
run;
```

Working Correlation Matrix
        Col1   Col2   Col3   Col4

```
Row1  1.0000  0.0000  0.0000  0.0000
Row2  0.0000  1.0000  0.0000  0.0000
Row3  0.0000  0.0000  1.0000  0.0000
Row4  0.0000  0.0000  0.0000  1.0000
```

GEE Fit Criteria
QIC 120.8579

Analysis Of GEE Parameter Estimates

Parameter		Estimate	Pr > |Z|
Intercept		51.5122	0.0122
gender	F	37.4045	<.0001
gender	M	0.0000	.
age		1.0692	0.0024
month		-1.0957	<.0001

The model with the Toeplitz correlation structure has the smallest QIC value and thus it is the optimal according to this criterion. All the predictors are significant, and the predicted model has the form $\widehat{E}(LDL) = 54.1516 + 36.0994 * female + 1.0174 * age - 0.8347 * month$. The estimated mean LDL level for females exceeds that for males by 37.5045 points. When age increases by one year, the estimated average cholesterol level increases by 1.0692 points. As estimated, the mean LDL level decreases by 1.0957 points for every one-month increase in time.

Predicted response at 3 months for a 48-year-old female patient is calculated as $LDL^0 = 54.1516 + 36.0994 + 1.0174 * 48 - 0.8347 * 3 = 136.5821$. SAS outputs the same predicted value:

```
data prediction;
input id gender$ age month;
cards;
28 F 48 3
;

data longform;
 set longform prediction;
run;

proc genmod;
class id gender;
 model ldl=gender age month/dist=normal link=identity;
```

```
  output out=outdata p=pred_ldl;
repeated subject=id/type=mdep(3) corrw;
run;

proc print data=outdata (firstobs=109 obs=109);
var pred_ldl;
run;
```

pred_ldl
136.580

The R script that fits the GEE models and the related output are given below.

```
cholesterol.data<- read.csv(file="./Example8.1Data.csv",
header=TRUE, sep=",")

#creating long-form data set
install.packages("reshape2")
library(reshape2)

longform.data<- melt(cholesterol.data,
id.vars=c("id", "gender", "age"),
variable.name = "LDLmonth", value.name="LDL")

#sorting long-form data set by id
longform.data<- longform.data[order(longform.data$id),]

#creating numeric variable for time
month<- ifelse(longform.data$LDLmonth=="LDL0", 0,
ifelse(longform.data$LDLmonth
=="LDL6", 6, ifelse(longform.data$LDLmonth=="LDL9",9,24)))

#specifying reference category
gender.rel<- relevel(cholesterol.data$gender, ref="M")

install.packages("geepack")
install.packages("MuMIn")
library(geepack)
library(MuMIn)

#fitting GEE model with unstructured working correlation matrix
summary(un.fitted.model<- geeglm(LDL ~ gender.rel + age + month,
```

```
data=longform.data, id=id,
family=gaussian(link="identity"), corstr="unstructured"))
QIC(un.fitted.model)
```

Coefficients:

	Estimate	Pr(>|W|)
(Intercept)	49.487	0.0061
gender.relF	34.315	1.3e-07
age	1.008	0.0006
month	-0.479	0.2396

Estimated Correlation Parameters:

	Estimate
alpha.1:2	1.138
alpha.1:3	0.644
alpha.1:4	0.142
alpha.2:3	0.608
alpha.2:4	0.153
alpha.3:4	0.427

QIC
656.1

Note that the estimated correlation matrix is not reliable because one of the estimates is larger than one. Therefore, we will omit this model from further consideration.

```
#fitting GEE model with autoregressive working correlation matrix
summary(ar1.fitted.model<- geeglm(LDL ~ gender.rel + age + month,
data=longform.data, id=id,
family=gaussian(link="identity"), corstr="ar1"))
QIC(ar1.fitted.model)
```

Coefficients:

	Estimate	Pr(>|W|)
(Intercept)	53.708	0.0042
gender.relF	36.463	1.5e-07
age	1.032	0.0012
month	-0.926	9.3e-08

Estimated Correlation Parameters:

	Estimate
alpha	0.701

QIC
650

```
#fitting GEE model with compound symmetric (exchangeable)
#working correlation matrix
summary(cs.fitted.model<- geeglm(LDL ~ gender.rel + age+ month,
data=longform.data, id=id,
family=gaussian(link="identity"), corstr="exchangeable"))
QIC(cs.fitted.model)
```

Coefficients:

	Estimate	Pr(>|W|)
(Intercept)	51.512	0.0122
gender.relF	37.405	3.0e-07
age	1.069	0.0024
month	-1.096	7.5e-09

Estimated Correlation Parameters:

	Estimate
alpha	0.582

QIC
651

```
#fitting GEE model with independent working correlation matrix
summary(ind.fitted.model<- geeglm(LDL ~ gender.rel + age + month,
data=longform.data, id=id,
family=gaussian(link="identity"), corstr="independence"))
QIC(ind.fitted.model)
```

Coefficients:

	Estimate	Pr(>|W|)
(Intercept)	51.512	0.0122
gender.relF	37.405	3.0e-07
age	1.069	0.0024
month	-1.096	7.5e-09

QIC
651

The QIC-optimal model is the one with autoregressive working correlation matrix. The statement below computes the predicted response based on this model.

```
#using AR fitted model for prediction
print(predict(ar1.fitted.model, data.frame(gender.rel="F",
age=48, month=3)))
```

136.9

□

Exercises for Chapter 8

EXERCISE 8.1. Show that in the random slope and intercept model defined in
(8.1),
(a) For any $i \neq i'$, $\mathbb{C}ov(y_{ij}, y_{i'j'}) = 0$. Note: j may be equal to j'.
(b) For any given i and $j \neq j'$, $\mathbb{C}ov(y_{ij}, y_{ij'}) = \sigma_{u_1}^2 + \sigma_{u_1 u_2}(t_j + t_{j'}) + \sigma_{u_2}^2 t_j t_{j'}$.
(c) The response variable y_{ij} has a normal distribution with the mean $\mathbb{E}(y_{ij}) = \beta_0 + \beta_1 x_{1ij} + \cdots + \beta_k x_{kij} + \beta_{k+1} t_j$ and variance $\mathbb{V}ar(y_{ij}) = \sigma_{u_1}^2 + 2\sigma_{u_1 u_2} t_j + \sigma_{u_2}^2 t_j^2 + \sigma^2$.

EXERCISE 8.2. A new manager in a department store wants to introduce a
new incentive compensation program for the floor employees. She looks at the
historical data on the yearly bonuses in the past three years, and such characteristics of each employee as the number of years with the company and the
current employment status (full- or part-time). The data for 15 employees are:

ID	Total Years	Status	Bonus in 2015	Bonus in 2016	Bonus in 2017
1	16	full	1482	1508	1543
2	7	part	673	710	895
3	11	full	933	1351	1440
4	8	part	844	958	1196
5	6	part	564	790	815
6	5	full	601	708	780
7	6	part	775	822	902
8	17	full	1209	1297	1475
9	12	full	929	1008	1255
10	9	full	983	1013	1111
11	11	full	909	1004	1084
12	6	part	387	853	999
13	4	part	476	530	627
14	6	full	780	843	925
15	10	full	717	1200	1399

(a) Carry out tests for normality of the bonus and plot the histogram. Is this variable normally distributed?

(b) Fit a random slope and intercept model regressing bonus (in thousands of dollars) on years with the company, status, and year (15, 16, or 17). Does the model fit the data well?

(c) Write down the fitted model, specifying all estimated parameters. What predictors are significant at the 5% significance level?

(d) Give interpretation of the estimated significant regression coefficients.

(e) According to the fitted model, what is the predicted bonus in 2017 for a full-time employee who has been with the company for 7 years?

EXERCISE 8.3. In order to improve the quality of medical service, an orthopedic clinic conducts a survey of their patients. After each visit, patients fill out a questionnaire, scoring the quality of service on the 0 to 10 continuous scale (larger values indicate better quality). The data for two doctors (referred to as A and B) are available. The variables are patient ID, gender, age (in years), doctor visited (A or B), length of the visit (in minutes) for three visits, and scores patients gave to these visits. The data on 35 patients are:

ID	Gender	Age	Doctor	Length1	Length2	Length3	Score1	Score2	Score3
101	F	78	A	25	20	25	7.1	7.5	7.6
102	F	63	A	30	30	40	5.5	5.8	6.1
103	F	62	A	10	15	10	10.0	10.0	9.8
104	F	71	B	15	15	40	7.8	7.3	7.5
105	M	68	A	40	60	40	3.5	3.5	3.0
106	F	63	A	25	15	20	8.5	8.7	8.8
107	F	60	B	25	35	25	6.7	5.7	6.5
108	F	70	A	20	20	20	9.0	8.3	8.2
109	F	57	A	30	20	15	8.4	7.8	8.1
110	F	59	B	25	30	15	7.1	7.4	7.9
111	M	62	A	50	30	70	3.0	3.2	2.6
112	M	58	A	20	15	45	6.1	6.8	6.9
113	M	75	A	25	35	30	5.7	5.6	4.7
114	M	76	B	35	50	25	4.9	5.4	5.2
115	F	75	A	15	20	25	8.2	8.9	8.2
116	M	57	A	45	30	40	4.6	3.9	3.2
117	F	68	A	35	25	40	3.8	4.8	5.3
118	M	65	B	40	40	25	3.9	3.9	4.7
119	F	67	B	20	15	30	6.5	7.2	6.6
120	F	60	B	25	15	15	7.3	7.1	7.8
121	F	67	A	15	20	15	7.7	8.0	8.3
122	F	57	B	10	15	15	9.8	9.2	8.6
123	M	62	B	55	60	75	3.4	2.7	2.3
124	M	71	A	20	30	25	7.1	6.6	7.4
125	M	71	B	15	15	20	8.8	9.1	9.3
126	M	64	A	25	30	30	5.6	6.3	6.3
127	M	51	A	35	40	30	5.1	4.6	3.9
128	F	70	B	35	25	15	6.8	7.1	7.6
129	M	61	A	35	40	50	5.5	5.2	4.8
130	M	62	B	60	40	65	3.7	3.4	2.4
131	F	68	A	20	35	35	5.3	5.6	4.9
132	F	68	B	35	30	15	7.2	6.2	5.6
133	M	64	B	40	20	30	5.4	4.9	4.5
134	F	76	B	30	45	25	5.5	4.7	4.6
135	F	78	B	25	20	15	7.6	8.3	9.2

(a) Create a long-form data set. Hint: When programming in R, use the code provided.

```
data1<- melt(quality.service.data
[,c("id", "gender", "age", "doctor","length1",
"length2", "length3")], id.vars=c("id","gender", "age", "doctor"),
variable.name="length.level", value.name="length")
data2<- melt(quality.service.data[,c("score1","score2", "score3")],
variable.name="score.level", value.name="score")
```

```
longform.data<- cbind(data1, data2)
```

(b) Confirm that the quality of service is normally distributed by plotting a histogram and conducting normality tests.

(c) Fit a random slope and intercept model to regress the quality of service scores on all the predictor variables. Discuss the model fit.

(d) What parameters of the random terms are significant at the 5% level? Are the scores for each patient correlated?

(e) What fixed-effects variables are significant predictors at the 5% significance level? Write down the fitted model.

(f) Interpret the estimates of the significant regression coefficients.

(g) Predict the quality of service score that would be given by a 55-year-old male who has a 30-minute appointment with Dr. A.

EXERCISE 8.4. Measurements were taken on 20 people involved in a physical fitness course. The data below contain participants' gender, age, oxygen intake (in ml per kg body weight per minute), run time (time to run 1 mile, in minutes), and pulse (average heart rate while running). The running was done under three different conditions: the first one on a treadmill, the second one on an indoor running track, and the third one on an outdoor running track. The data are summarized in the table below.

ID	Gender	Age	Oxgn1	Rtime1	Pulse1	Oxgn2	Rtime2	Pulse2	Oxgn3	Rtime3	Pulse3
1	F	39	44.6	11.4	165	39.8	10.3	166	36.7	11.1	162
2	F	32	45.3	10.1	181	40.6	12.3	180	40.2	10.8	180
3	F	53	54.3	12.7	163	59.0	17.8	165	58.2	15.4	158
4	F	33	44.8	11.6	168	42.2	12.3	170	38.8	10.4	165
5	M	45	39.4	13.1	162	39.5	11.3	153	39.5	11.2	143
6	F	34	60.1	9.6	160	55.1	9.8	167	49.0	9.4	154
7	F	47	50.5	12.1	162	53.4	12.1	170	50.4	12.8	161
8	M	50	37.4	14.0	176	39.4	14.7	176	39.0	14.0	173
9	M	49	44.8	13.1	165	40.9	14.1	163	38.5	12.7	167
10	M	37	51.9	10.3	155	48.6	10.8	157	57.4	9.8	151
11	M	34	49.2	9.0	174	43.9	10.4	180	44.7	10.5	170
12	M	44	40.8	11.0	153	40.9	12.1	159	36.1	11.5	158
13	F	33	50.4	10.1	155	40.1	9.7	170	41.1	9.0	150
14	F	50	39.4	12.6	165	36.6	11.9	154	37.0	11.9	160
15	F	46	46.1	11.2	162	45.9	12.4	170	42.6	12.7	152
16	M	35	45.4	9.6	151	40.3	11.0	157	41.1	9.6	148
17	M	48	39.2	12.9	159	40.6	12.7	165	37.4	12.1	142
18	M	36	45.8	10.5	175	42.3	9.6	173	44.8	10.5	173
19	M	39	48.7	9.4	172	38.7	11.8	176	39.7	10.6	172
20	M	42	47.9	11.5	161	39.3	11.2	166	44.3	11.6	153

(a) Check that pulse has normal distribution. Construct a histogram and conduct normality tests.

(b) Run a random slope and intercept regression model for Pulse. Discuss the model fit.

(c) Specify the fitted model. What parameters of the random-effects terms are significant at the 5% level? What fixed-effects terms are significant at the 5% level?

(d) Interpret the estimated regression coefficients for the significant fixed-effects terms.

(e) Predict an average heart rate for a 36-year-old woman who is running on a treadmill, if her oxygen intake is 40.2 units, and her run time is 10.3 minutes per mile.

EXERCISE 8.5. A health center conducted a study on efficacy of an intervention on weight loss. The intervention consisted of a lecture on proper nutrition and importance of exercising, followed by a cooking class. The study had a wait list control group. For each of the 34 study participants, the investigators recorded the group (intervention or control), gender (F/M), the typical length of daily exercise in the past week (in minutes), and BMI (in kg/m^2) at the beginning of the study, and at 1 and 3 months afterwards. The data are as follows:

ID	Group	Gender	aExercise	aBMI	bExercise	bBMI	cExercise	cBMI
1	Int	F	0	42.4	150	40.0	150	36.8
2	Int	F	15	32.9	220	30.6	225	28.6
3	Int	M	10	32.0	300	30.8	300	26.1
4	Int	M	20	26.1	180	25.5	180	21.1
5	Int	F	0	27.5	200	26.4	220	22.5
6	Int	F	30	40.4	175	38.3	180	32.1
7	Int	M	15	33.5	350	28.2	350	25.8
8	Int	F	15	35.2	185	34.8	190	30.6
9	Int	F	0	39.5	155	37.1	150	35.3
10	Int	M	20	27.3	300	26.3	310	22.6
11	Int	M	0	46.9	150	43.5	150	40.3
12	Int	M	20	34.4	180	32.2	185	28.1
13	Int	F	0	34.2	165	31.0	165	26.8
14	Int	F	45	26.5	300	24.6	330	20.8
15	Int	F	0	29.6	200	28.2	200	24.9
16	Int	F	10	31.2	180	29.3	190	28.6
17	Cnt	F	40	29.3	25	28.9	30	26.3
18	Cnt	M	20	45.9	10	43.1	15	42.9
19	Cnt	M	40	41.5	20	38.8	30	39.9
20	Cnt	F	30	33.3	25	33.4	35	33.2
21	Cnt	M	15	31.1	35	30.9	0	30.9
22	Cnt	F	10	43.3	35	43.6	30	44.5
23	Cnt	M	15	35.5	0	36.5	50	35.3
24	Cnt	F	10	42.4	15	43.4	50	42.3
25	Cnt	F	20	37.0	30	36.6	45	35.5
26	Cnt	M	0	37.8	30	35.7	45	34.3
27	Cnt	F	20	23.7	10	23.1	0	23.7
28	Cnt	F	10	38.7	15	20.4	25	20.1
29	Cnt	F	0	41.2	15	41.2	55	39.7
30	Cnt	F	30	30.2	35	29.9	5	29.4
31	Cnt	M	10	38.4	20	38.1	30	37.0
32	Cnt	F	10	37.5	15	37.4	5	36.8
33	Cnt	M	30	34.5	10	34.4	20	33.9
34	Cnt	M	15	37.6	35	36.2	25	36.0

(a) Verify normality of the response variable BMI by plotting the histogram and carrying out normality tests.

(b) Fit the random slope and intercept model. How good is the model fit?

(c) Present the fitted model and specify all estimated parameters. Discuss significance of the parameters at the 5% significance level.

(d) Give interpretation of the estimated significant beta coefficients. Is the intervention efficient?

(e) Compute the predicted BMI at 3 months for an intervention group female participant, if she exercises for 1 hour every day.

EXERCISE 8.6. For the data in Exercise 8.3,
(a) Fit random slope and intercept models with unstructured, Toeplitz, spatial power, autoregressive, and compound symmetric covariance matrices for error terms, whichever converge.
(b) Which of the fitted models has the best fit according to the AIC, AICC, and BIC criteria?
(c) Answer parts (e)-(g) in Exercise 8.3 as applied to the best fitted model.

EXERCISE 8.7. Consider the data in Exercise 8.2. Answer the questions below.
(a) Fit random slope and intercept models with unstructured, Toeplitz, spatial power, autoregressive, and compound symmetric covariance matrices for error terms. Present the results for those models that converge.
(b) Discuss the fit of the models with respect to the AIC, AICC, and BIC criteria. Which model is optimal?
(c) For the optimal model, answer questions (c)-(e) in Exercise 8.2.

EXERCISE 8.8. Use the data in Exercise 8.4 to do the following analysis:
(a) Run random slope and intercept regressions to model the pulse. Try unstructured, Toeplitz, spatial power, autoregressive, and compound symmetric covariance structures for the error terms.
(b) Find the best-fitted model according to the AIC, AICC, and BIC criteria.
(c) Answer questions (c)-(e) in Exercise 8.4 for the model that fits the data the best.

EXERCISE 8.9. Take the data presented in Exercise 8.5.
(a) For BMI, fit the random slope and intercept regression models with unstructured, Toeplitz, spatial power, autoregressive, and compound symmetric covariance structures for the error terms.
(b) Which of the models has the best fit according to AIC, AICC, and BIC criteria?
(c) For the best-fitted model, do the analysis for questions (c) through (e) in Exercise 8.5.

EXERCISE 8.10. For the data in Exercise 8.3,

(a) Fit the Generalized Estimating Equations models with unstructured, Toeplitz, autoregressive, compound symmetric, and independent working correlation matrices.

(b) Which of the fitted models has the best fit according to the QIC criterion?

(c) Answer parts (e)-(g) in Exercise 8.3 in relation to the best-fitted model.

EXERCISE 8.11. Returning to the data in Exercise 8.2, answer the following questions:

(a) Fit the GEE models with unstructured, Toeplitz, autoregressive, compound symmetric, and independent working correlation matrices.

(b) Find the best model using the QIC criterion.

(c) For the model that fits the data the best, answer questions (c)-(e) in Exercise 8.2.

EXERCISE 8.12. Consider the data in Exercise 8.4.

(a) Run the Generalized Estimating Equations models with unstructured, Toeplitz, autoregressive, compound symmetric, and independent working correlation matrices for the pulse.

(b) Compare the QIC values for the fitted models and choose the optimal one.

(c) For the optimal model, do questions (c)-(e) in Exercise 8.4.

EXERCISE 8.13. For the data given in Exercise 8.5, do the following questions:

(a) Fit the GEE models with unstructured, Toeplitz, autoregressive, compound symmetric, and independent working correlation matrices of the response variable BMI.

(b) Choose the best-fitted model with respect to the QIC criterion.

(c) For the best-fitted model, do parts (c)-(e) in Exercise 8.5.

Chapter 9

Generalized Linear Regression Models for Repeated Measures Data

Generalized linear mixed-effects and GEE models are regression models that extend the classes of random slope and intercept (Section 8.1) and GEE (Section 8.3) models for normally distributed response variables, respectively. Generalized models can be used to model repeated measures data with skewed, categorical, or count response.

Random slope and intercept models with covariance structure for errors that we considered in Section 8.2 are not applicable in this case, because the models for data with non-normally distributed response don't contain error terms.

9.1 Generalized Random Slope and Intercept Model

9.1.1 Model Definition

A *generalized random slope and intercept model* (also termed the *generalized linear mixed model*) is a generalization of the linear mixed-effects model introduced in (8.1) to any distribution of the response variable in the exponential family of distributions, and the corresponding link function. In this chapter we talk about the models considered earlier: gamma (in Section 2.2), binary logistic (in Section 3.1), Poisson (in Section 5.1), negative binomial (in Section 6.1), and beta (in Section 7.1). In the setting of each regression, we assume that the data were collected longitudinally at times $t_1, \ldots, t_p$, and that for the i-th individual at time t_j the observations are $x_{1ij}, \ldots, x_{kij}$, and y_{ij} where $i = 1, \ldots, n$,

257

and $j = 1, ..., p$, and y_{ij}'s have one of the aforementioned distributions. The generalized mixed-effects model in each case has two random additive terms in the linear regression expression: $\beta_0 + \beta_1\, x_{1ij} + \cdots + \beta_k\, x_{kij} + \beta_{k+1}\, t_j + u_{1i} + u_{2i}\, t_j$ where u_{1i}'s are independent $\mathcal{N}(0, \sigma_{u_1}^2)$ random intercepts, u_{2i}'s are independent $\mathcal{N}(0, \sigma_{u_2}^2)$ random slopes, $\mathbb{C}ov(u_{1i}, u_{2i}) = \sigma_{u_1 u_2}$, and $\mathbb{C}ov(u_{1i}, u_{2i'}) = 0$ for $i \neq i'$.

9.1.2 Fitted Model, Interpretation of Estimated Regression Coefficients, Model Goodness-of-Fit Check, and Predicted Response

Fitted generalized random slope and intercept model has the form

$$g\big(\widehat{\mathbb{E}}(y)\big) = \widehat{\beta}_0 + \widehat{\beta}_1\, x_1 + \cdots + \widehat{\beta}_k\, x_k + \widehat{\beta}_{k+1}\, t$$

where $g(\cdot)$ is the link function that corresponds to the underlying distribution.

Interpretation of the estimated beta coefficients and computation of predicted response for a fixed set of predictors are done the same way as in the appropriate regression model (gamma, binary logistic, Poisson, negative binomial, or beta).

The goodness-of-fit test is the standard deviance test where the null model contains only fixed-effects terms.

9.1.3 SAS Implementation

The procedure `glimmix`, employed in Section 7.1 for fitting a beta regression, can handle generalized linear mixed-effects models if the `random` statement is added. The general statements should be:

```
proc glimmix data=data_name method=Laplace;
class catpredictor1_name (ref='level_name') catpredictor2_name
        (ref='level_name') ... ;
model response_name=<list of predictors>/dist=dist_name
  link=link_type solution;
    output out=outdata pred(ilink)=predicted_name;
      random intercept time_name/subject=id_name type=un;
    covtest/wald;
run;
```

- The option `method=Laplace` specifies the method of parameter estimation, in which the output contains the values of -2*log-likelihood. This value is needed for computation of model deviance and carrying out the test of model fit.
- To model level '1' in logistic regression, include `(event='1')` after *response_name* in the `model` statement.
- The choices for the distributions and corresponding links are: `dist=gamma link=log`, `dist=binomial link=logit`, `dist=poisson link=log`, `dist=nb link=log`, and `dist=beta link=logit`.
- Along with specific values for all predictors, the row for prediction must contain a hypothesized value for *id_name*.
- The statement `covtest/wald` should be added to request the Wald z-test results for the parameters of the random-effects terms.

9.1.4   R Implementation

To fit a generalized random slope and intercept model in R, the function `glmer()` in the library `lme4` is used. The data set must be in the long form. The general script is:

summary(*fitted.model.name*<- glmer(*response.name* ~ *x1.name* + ···
+ *xk.name* + *time.name* + (1 + *time.name*|*id.name*),
data=*longform.data.name*,
family=*dist.name*(link="*link.type*")))

- Currently there is no built-in procedure to carry out tests of significance for the variance-covariance parameters, but R outputs the estimates and their standard errors. Thus, the two-sided z-test may be conducted by hand.
- The distribution names and the link types can be specified as `gamma(link="log")`, `binomial(link="logit")`, `poisson(link="log")`, `nb(link="log"))`, or `beta(link="logit")`.
- To output the predicted value of the response variable for a fixed set of predictors, the function `predict()` should be used with the following syntax:

predict(*fitted.model.name*, data.frame(*catpredictor1.name*="*value*", ... ,
numpredictork.name=*value*, *time.name*=*value*), re.form=NA,
type="response")

The argument `re.form=NA` requests predicted response not conditioned on any values of the random-effects terms.

9.1.5   Example

EXAMPLE 9.1. A dermatologist tests a new ointment treatment for psoriasis. He administers the ointment to five patients, and keeps five patients as control. The control patients take a medication that is commonly prescribed against the disease. The doctor sees the patients next day, then after one week, two weeks, five weeks, and finally, after three months. He records the number of psoriatic patches that are visible on patients' bodies. The code below fits the random slope and intercept Poisson model to the data.

```
data psoriasis;
input patid group$ day1 week1 week2 week5 month3 @@;
cards;
1 Tx 15 12 9  3  0    2 Tx 24 17 9  8  8
3 Cx 14 14 15 15 14   4 Cx 23 20 19 19 15
5 Tx 11 10 3  2  0    6 Cx 11 10 8  7  5
7 Tx 7  6  5  3  0    8 Tx 9  6  2  0  0
9 Cx 9  9  9  9  9    10 Cx 21 16 16 15 15
;

data longform;
 set psoriasis;
  array w{5}(0.14 1 2 5 13);
  array x{5} day1 week1 week2 week5 month3;
      do i=1 to 5;
        weeks=w{i};
         npatches=x{i};
         output;
      end;
    keep patid group weeks npatches;
run;

proc glimmix method=Laplace;
 class group;
  model npatches=group weeks/solution
    dist=poisson link=log;
 random intercept weeks/subject=patid type=un;
   covtest/wald;
run;
```

Covariance Parameter Estimates
Cov Parm   Estimate   Pr Z

260

UN(1,1)	0.1139	0.1292
UN(2,1)	0.04259	0.2046
UN(2,2)	0.02408	0.0749

Solutions for Fixed Effects

Effect	group	Estimate	Pr > |t|
Intercept		2.6503	<.0001
group	Cx	-0.1916	0.5762
group	Tx	0	.
weeks		-0.1570	0.0229

The output shows that the variances and covariance of the random-effects terms are not significant at the 5% level. A reasonable simpler model to run in this case would be a random intercept-only model, where the random slope is omitted.

```
proc glimmix method=Laplace;
  class group;
   model npatches=group weeks/solution
     dist=poisson link=log;
   random intercept/subject=patid type=un;
     covtest/wald;
run;
```

-2 Log Likelihood 289.31

Covariance Parameter Estimates

Cov Parm	Estimate	Pr Z
UN(1,1)	0.1492	0.0281

Solutions for Fixed Effects

Effect	group	Estimate	Pr > |t|
Intercept		2.0222	<.0001
group	Cx	0.7391	0.0080
group	Tx	0	.
weeks		-0.05778	<.0001

In the fitted random intercept-only Poisson model, the estimated rate $\widehat{\lambda} = \widehat{E}(y) = \exp\{2.0222 + 0.7391 * Cx - 0.05778 * weeks\}$. Both group and weeks are significant predictors at the 0.05 level. The estimated average number of psoriatic patches for patients in the control group is $\exp\{0.7391\} \cdot 100\% = 209.41\%$ of that for the patients in the treatment group. Each week, the estimated average number of patches changes by $(\exp\{-0.05778\} - 1) \cdot 100\% = -5.61\%$, that is, decreases by 5.61%.

Next, we fit the null model, which in this case is the model with only the fixed effects, and conduct the deviance test of model fit.

```
proc glimmix;
class group;
  model npatches=group weeks/solution
    dist=poisson link=log;
run;

data deviance_test;
  deviance=336.27-289.31;
    pvalue=1-probchi(deviance,1);
run;

proc print;
run;
```

deviance   pvalue
  46.96   7.245E-12

The p-value is tiny which indicates a good model fit. Finally, we would like to use the estimated model to predict the number of psoriatic patches for a patient in the treatment group at five weeks. We compute $y^0 = \exp\{2.0222 - 0.05778 * 5\} = 5.66$. To produce the prediction in SAS, we write:

```
data prediction;
input patid group$ weeks;
cards;
11 Tx 5
;

data longform;
  set longform prediction;
run;

proc glimmix method=Laplace;
  class group;
    model npatches=group weeks/solution
      dist=poisson link=log;
```

```
  random intercept/subject=patid type=un;
  output out=outdata pred(ilink)=p_npatches;
run;

proc print data=outdata (firstobs=51 obs=51);
var p_npatches;
run;
```

p_npatches
   5.65951

Note that the row for prediction must contain a (hypothesized) value of the id variable.

Further, the R script and appropriate output for this example are as follows:

```
psoriasis.data<-read.csv(file="./Example9.1Data.csv",
header=TRUE, sep=",")

#creating long-form data set
install.packages("reshape2")
library(reshape2)

longform.data<- melt(psoriasis.data, id.vars=c("patid", "group"),
variable.name = "visits", value.name="npatches")

#creating numeric variable for time
weeks<- ifelse(longform.data$visits=="day1", 0.14,
ifelse(longform.data$visits =="week1", 1,
ifelse(longform.data$visits=="week2",
2,ifelse(longform.data$visits=="week5",
5, 13))))

#specifying reference category
group.rel<- relevel(longform.data$group, ref="Tx")

#fitting random slope and intercept Poisson model
install.packages("lme4")
library(lme4)

summary(glmer(npatches ~ group.rel + weeks
+ (1 + weeks|patid), data=longform.data,
family=poisson(link="log")))
```

Random effects:
 Groups Name   Variance Corr
 patid (Intercept) 0.11393
        weeks   0.02403 0.81

Fixed effects:
           Estimate Pr(>|z|)
(Intercept) 2.65008 < 2e-16
group.relCx -0.19148 0.57023
weeks   -0.15681 0.00565

```
#fitting random intercept-only Poisson model
summary(fitted.model<- glmer(npatches ~ group.rel
+ weeks + (1|patid),
data=longform.data, family=poisson(link="log")))
```

  AIC   BIC
297.3   305.0

Random effects:
 Groups Name   Variance
 patid (Intercept) 0.1492

Fixed effects:
           Estimate Pr(>|z|)
(Intercept) 2.02224 < 2e-16
group.relCx 0.73904   0.00515
weeks   -0.05778 2.74e-07

```
#computing AICC
n<- 50
p<- 4
print(AICC<- -2*logLik(fitted.model)+2*p*n/(n-p-1))
```

298.197

```
#checking model fit
null.model<- glm(npatches ~ group.rel + weeks,
data=longform.data, family=poisson(link=log))
print(deviance<- -2*(logLik(null.model)-logLik(fitted.model)))
```

46.96343

```
print(p.value<- pchisq(deviance, df=1, lower.tail=FALSE))
```

7.232377e-12

```
#using the model for prediction
print(predict(fitted.model, data.frame(patid=11, group.rel="Tx",
weeks=5), re.form=NA, type="response"))
```

5.659691

□

9.2   Generalized Estimating Equations Model

9.2.1   Model Definition

The Generalized Estimating Equations model described in Section 8.3 translates directly to the gamma, logistic, Poisson, and negative binomial models. Neither SAS nor R can handle GEE for a beta distribution, so this model will be omitted from consideration.

9.2.2   SAS Implementation

In SAS, `proc genmod` is used to fit GEEs with the syntax given in Subsection 8.3.4. Allowable distributions and link types are: `dist=gamma link=log`, `dist=bin link=logit`, `dist=poisson link=log`, and `dist=nb link=log`. When the distribution is logistic, include `descending` in the *proc* statement to model the probability of level '1'.

9.2.3   R Implementation

In R, the function `geeglm()` is employed for fitting a Generalized Estimating Equations model. The general form of the script was presented in Subsection 8.3.5.

• The underlying response distributions and the corresponding links are given as: `gamma(link="log")`, `binomial(link="logit")`, `poisson(link="log")`, and `nb(link="log")`.

• To request predicted response, the argument `type="response"` should be added to the function `predict()`.

9.2.4 Example

EXAMPLE 9.2. We fit GEE models to the data in Example 9.1. In SAS, the model with unstructured working correlation matrix doesn't converge. Therefore, we present results for the Toeplitz, autoregressive, exchangeable, and independent structures.

```
/*fitting GEE with Toeplitz working
 correlation matrix*/
proc genmod;
   class patid group;
     model npatches=group weeks/dist=poisson link=log;
   repeated subject=patid/type=mdep(3) corrw;
 run;
```

Working Correlation Matrix

	Col1	Col2	Col3	Col4	Col5
Row1	1.0000	0.7495	0.4228	0.2298	0.0000
Row2	0.7495	1.0000	0.7495	0.4228	0.2298
Row3	0.4228	0.7495	1.0000	0.7495	0.4228
Row4	0.2298	0.4228	0.7495	1.0000	0.7495
Row5	0.0000	0.2298	0.4228	0.7495	1.0000

GEE Fit Criteria
QIC -483.3303

Analysis Of GEE Parameter Estimates

Parameter		Estimate	Pr > |Z|
Intercept		2.1455	<.0001
group	Cx	0.6625	0.0303
group	Tx	0.0000	.
weeks		-0.0362	0.0019

```
/*fitting GEE with autoregressive working
 correlation matrix*/
proc genmod;
   class patid group;
     model npatches=group weeks/dist=poisson link=log;
   repeated subject=patid/type=ar corrw;
 run;
```

Working Correlation Matrix

	Col1	Col2	Col3	Col4	Col5
Row1	1.0000	0.7504	0.5631	0.4225	0.3171
Row2	0.7504	1.0000	0.7504	0.5631	0.4225
Row3	0.5631	0.7504	1.0000	0.7504	0.5631
Row4	0.4225	0.5631	0.7504	1.0000	0.7504
Row5	0.3171	0.4225	0.5631	0.7504	1.0000

GEE Fit Criteria
QIC -494.0739

Analysis Of GEE Parameter Estimates

Parameter		Estimate	Pr > |Z|
Intercept		2.2059	<.0001
group	Cx	0.5637	0.0379
group	Tx	0.0000	.
weeks		-0.0424	0.0016

```
/*fitting GEE with exchangeable working
correlation matrix*/
proc genmod;
   class patid group;
     model npatches=group weeks/dist=poisson link=log;
   repeated subject=patid/type=cs corrw;
 run;
```

Working Correlation Matrix

	Col1	Col2	Col3	Col4	Col5
Row1	1.0000	0.5775	0.5775	0.5775	0.5775
Row2	0.5775	1.0000	0.5775	0.5775	0.5775
Row3	0.5775	0.5775	1.0000	0.5775	0.5775
Row4	0.5775	0.5775	0.5775	1.0000	0.5775
Row5	0.5775	0.5775	0.5775	0.5775	1.0000

GEE Fit Criteria
QIC -438.9074

Analysis Of GEE Parameter Estimates

Parameter		Estimate	Pr > |Z|
Intercept		2.3502	<.0001
group	Cx	0.3174	0.1689
group	Tx	0.0000	.
weeks		-0.0579	0.0054

```
/*fitting GEE with independent working
 correlation matrix*/
proc genmod;
   class patid group;
     model npatches=group weeks/dist=poisson link=log;
   repeated subject=patid/type=ind corrw;
 run;
```

Working Correlation Matrix

	Col1	Col2	Col3	Col4	Col5
Row1	1.0000	0.0000	0.0000	0.0000	0.0000
Row2	0.0000	1.0000	0.0000	0.0000	0.0000
Row3	0.0000	0.0000	1.0000	0.0000	0.0000
Row4	0.0000	0.0000	0.0000	1.0000	0.0000
Row5	0.0000	0.0000	0.0000	0.0000	1.0000

GEE Fit Criteria
QIC -459.7895

Analysis Of GEE Parameter Estimates

Parameter		Estimate	Pr > |Z|
Intercept		2.1223	<.0001
group	Cx	0.6902	0.0118
group	Tx	0.0000	.
weeks		-0.0578	0.0052

The model with the autoregressive working correlation structure has the smallest QIC value, thus, this model has the best fit. In this case, the estimated rate $\widehat{\lambda} = \widehat{E}(y) = \exp\{2.2059 + 0.5637 * Cx - 0.0424 * weeks\}$, where both predictors are significant at the 5% significance level. The estimated average number of psoriatic patches for patients in the control group is $\exp\{0.6902\} \cdot 100\% = 199.41\%$ of that for the patients in the treatment group. Each week, the estimated average number of patches changes by $(\exp\{-0.0578\} - 1) \cdot 100\% = -5.62\%$, that is, decreases by 5.62%.

To predict the number of psoriatic patches for a patient in the treatment group five weeks into the study, we calculate $y^0 = \exp\{2.2059 - 0.0424 * 5\} = 7.34$. In SAS, the following statements produce the prediction:

```
data prediction;
input patid group$ weeks;
cards;
```

```
11 Tx 5
;

data longform;
 set longform prediction;
run;

proc genmod;
 class patid group;
  model npatches=group weeks/dist=poisson link=log;
    output out=outdata p=p_npatches;
  repeated subject=patid/type=ar corrw;
 run;

proc print data=outdata (firstobs=51 obs=51);
var p_npatches;
run;

p_npatches
   7.34378
```

Below we give the R statements for this example and corresponding outputs. We use the long-form data set already created and sorted in Example 9.1.

```
#specifying reference category
group.rel<- relevel(longform.data$group, ref="Tx")

#fitting GEE model with autoregressive working correlation matrix
install.packages("geepack")
install.packages("MuMIn")
library(geepack)
library(MuMIn)

summary(ar1.fitted.model<- geeglm(npatches ~ group.rel + weeks,
data=longform.data, id=patid,
family=poisson(link="log"), corstr="ar1"))
QIC(ar1.fitted.model)
```

Coefficients:

	Estimate	Pr(>|W|)
(Intercept)	2.18579	< 2e-16
group.relCx	0.59536	0.03025
weeks	-0.04452	0.00183

Estimated Correlation Parameters:
    Estimate
alpha   0.6773

  QIC
-1404

```
#fitting GEE model with exchangeable working correlation matrix
summary(exch.fitted.model<- geeglm(npatches ~ group.rel + weeks,
data=longform.data, id=patid, family=poisson(link="log"),
corstr="exchangeable"))
QIC(exch.fitted.model)
```

Coefficients:
          Estimate Pr(>|W|)
(Intercept) 2.3890   <2e-16
group.relCx 0.2476   0.2732
weeks   -0.0580   0.0055

Estimated Correlation Parameters:
    Estimate
alpha   0.625

  QIC
-1403

```
#fitting GEE model with independent working correlation matrix
summary(ind.fitted.model<- geeglm(npatches ~ group.rel + weeks,
data=longform.data, id=patid, family=poisson(link="log"),
corstr="independence"))
QIC(ind.fitted.model)
```

Coefficients:
          Estimate Pr(>|W|)
(Intercept) 2.1223   <2e-16
group.relCx 0.6902   0.0118
weeks   -0.0578   0.0052

Estimated Correlation Parameters:
    Estimate
alpha   0.625

  QIC
-1403

The model with autoregressive working correlation matrix has the smallest QIC value, thus it has the best fit. Below we use this model for prediction.

```
#using AR fitted model for prediction
print(predict(ar1.fitted.model, type="response",
data.frame(patid=11, group.rel="Tx", weeks=5)))
```

7.122

□

Exercises for Chapter 9

EXERCISE 9.1. A center for medical weight loss is considering buying a shipment of a new medication that was recently marketed. Prior to the purchase, the doctors decide to conduct a trial with their patients to test superiority of this new medication as compared to the one that they have been prescribing on the regular basis. They administer this medication to 8 patients while keeping another 8 patients as a control group. For each subject in the trial, the doctors record group (Tx/Cx), gender (M/F), and excess body weight loss (EWL, in percent) between measurement sessions: from the baseline to 1 month (EWL 1), 1 month to 2 months (EWL 2), 2 months to 3 months (EWL 3), and 3 months to 6 months (EWL 4). The data are:

Patid	Group	Gender	EWL1	EWL2	EWL3	EWL 4
1	Tx	M	11.8	16.4	7.1	4.5
2	Tx	F	18.3	7.7	10.7	4.1
3	Tx	F	20.1	8.2	7.2	6.3
4	Tx	F	15.6	7.8	7.2	2.7
5	Tx	M	12.5	8.6	9.7	5.4
6	Tx	F	24.4	8.7	6.6	4.7
7	Tx	F	18.8	12.3	6.7	4.5
8	Tx	M	11.2	9.1	5.6	3.1
9	Cx	F	13.9	14.3	4.1	5.0
10	Cx	F	6.8	5.2	4.5	1.4
11	Cx	M	8.1	12.7	12.3	4.9
12	Cx	F	5.6	16.5	4.8	1.8
13	Cx	M	9.6	9.9	3.6	3.5
14	Cx	M	6.8	7.5	5.1	1.7
15	Cx	F	4.7	8.3	3.2	2.4
16	Cx	F	6.7	4.1	2.4	1.3

(a) Plot the histogram of the percent excess body weight loss to see that this variable has a distribution with a long right tail.

(b) Try to run a generalized random slope and intercept model based on a gamma distribution. If it doesn't run, fit an intercept-only model. Discuss the fit of this model. Hint: as time variable, use visits with values 1, 2, 3, or 4.

(c) What parameters are significant at the 5% level? Give the fitted model, specifying all parameter estimates.

(d) Give interpretation of the estimates of significant fixed-effects parameters. Is the new medication superior to the regularly used one?

(e) What percent excess body weight loss can the doctors expect to see between 3 and 6 months in their males patients who will be taking this new medication?

EXERCISE 9.2. A graduate student in Sport and Exercise Nutrition Science recruited 50 freshmen for a study for his Master's thesis. At the beginning of the study he gave a 30-minute presentation on the necessity of fruit consumption and exercising. The participants were then asked to record every day the number of fruits they eat and the number of minutes they exercise (rounded to the nearest five minutes). In the intervention group, the participants were also asked to send this information to the graduate student by a daily text message, while in the control group, no messaging took place. For the analysis, the graduate student chose fruit consumption and daily exercise at day 1, day 7, day 14, and day 30. Seven participants dropped out of the study early, so no data were recorded for them. The complete data on 43 participants are presented in the table below.

ID	Group	Gender	aFruits	aExercise	bFruits	bExercise	cFruits	cExercise	dFruits	dExercise
1	Tx	M	0	0	4	75	5	15	4	30
2	Tx	F	0	0	2	75	4	60	3	60
3	Tx	F	2	0	5	60	6	60	7	45
4	Tx	F	1	15	4	75	5	75	4	45
5	Tx	F	1	0	2	60	5	45	5	75
6	Tx	F	1	0	3	45	4	30	4	60
7	Tx	F	0	0	3	75	5	60	5	75
8	Tx	F	0	30	4	60	4	60	2	30
9	Tx	F	0	30	4	75	5	75	5	30
10	Tx	M	1	0	2	60	4	30	6	15
11	Tx	F	2	0	3	60	3	75	3	75
12	Tx	M	0	30	3	75	4	75	4	15
13	Tx	F	1	30	5	30	4	30	6	60
14	Tx	M	0	30	4	45	4	60	3	30
15	Tx	M	1	30	5	60	3	60	2	45
16	Tx	M	0	15	4	30	3	60	2	60
17	Tx	M	1	15	5	30	5	30	6	75
18	Tx	M	0	0	3	15	3	75	4	60
19	Tx	M	2	30	2	45	5	30	2	30
20	Tx	M	2	30	2	45	5	45	5	75
21	Tx	F	0	15	8	60	8	45	12	60
22	Tx	M	0	15	3	60	3	15	3	45
23	Tx	M	1	0	4	15	5	45	5	30
24	Cx	M	0	0	1	15	0	15	2	15
25	Cx	M	0	0	2	15	0	30	2	30
26	Cx	F	1	0	2	15	1	45	1	30
27	Cx	F	1	15	0	0	3	0	1	15
28	Cx	M	2	15	0	15	2	30	2	30
29	Cx	M	2	0	2	0	0	0	3	45
30	Cx	F	1	0	0	0	3	30	3	15
31	Cx	F	0	15	0	30	1	15	2	15
32	Cx	F	2	30	1	30	1	30	3	45
33	Cx	M	1	0	1	30	2	30	1	30
34	Cx	M	2	30	1	0	2	30	0	15
35	Cx	M	2	0	1	0	2	0	1	30
36	Cx	M	1	15	2	30	1	45	1	30
37	Cx	F	1	15	1	30	1	15	2	15
38	Cx	M	1	30	0	15	2	30	2	30
39	Cx	F	2	30	2	0	3	15	0	45
40	Cx	M	1	0	0	30	1	45	2	45
41	Cx	F	1	0	1	15	0	15	0	45
42	Cx	F	2	0	2	15	3	15	3	45
43	Cx	M	1	0	2	15	0	30	1	15

(a) Construct a histogram for the amount of daily exercise. Show that the density is right-skewed.

(b) Run a generalized random slope and intercept model with the gamma distribution of the response. If it doesn't run, reduce to random intercept-only model. How well does the model fit the data?

(c) Write down the fitted model. Discuss significance of its parameters at the 5% level.

(d) Does the intervention appear to be working? Interpret the estimates of all significant regression coefficients.

(e) Find the predicted number of minutes of exercising for a female student in the intervention group on the 30th day.

EXERCISE 9.3. A pharmaceutical company conducted a dosage trial for a painkiller medication. Five dosages were tested and two were picked as the winners (dosages A and B). Further investigation was launched into a long-run safety of these dosages. An experiment was set up with 14 subjects in each of the two groups, taking dosages A and B, respectively. Equal number of men and women were assigned to each group. The goal of the experiment was to identify which dosage is less likely to cause side effects. The data set below contains participants' IDs, dosage (A or B), gender, and presence or absence of side effects (1=present, or 0=absent) at 1, 3, 7, and 16 weeks.

ID	Dosage	Gender	Week1	Week3	Week7	Week16
1	A	F	0	0	1	1
2	A	F	0	0	1	1
3	A	F	0	0	0	1
4	A	F	0	0	1	1
5	A	F	0	0	0	1
6	A	F	0	0	0	0
7	A	F	0	0	0	1
8	A	M	0	0	1	1
9	A	M	0	0	0	0
10	A	M	0	0	0	0
11	A	M	0	0	0	1
12	A	M	0	0	0	1
13	A	M	0	0	0	1
14	A	M	0	0	1	1
15	B	F	0	0	0	1
16	B	F	0	1	1	1
17	B	F	0	0	0	1
18	B	F	0	0	1	1
19	B	F	1	1	1	1
20	B	F	0	0	1	1
21	B	F	0	0	1	1
22	B	M	0	0	0	1
23	B	M	1	1	1	1
24	B	M	0	0	1	1
25	B	M	1	1	1	1
26	B	M	1	1	1	1
27	B	M	0	1	1	1
28	B	M	1	1	1	1

(a) Model the logistically distributed presence of side effects via a generalized random slope and intercept model. If it doesn't converge, then run the random intercept-only model. Discuss the model fit.

(b) Specify the fitted model, giving the estimates of all parameters. Which parameters are significant at the 5% significance level?

(c) Interpret the estimates of the significant beta coefficients. What dosage should be preferred?

(d) Predict the odds in favor of a side effect occurring at week 7 for a woman taking dosage A.

EXERCISE 9.4. Use the data given in Exercise 9.2 to model the daily consumption of fruits.

(a) Fit a generalized random slope and intercept model (or, if fails to converge, then fit random intercept-only model) to model daily consumption of fruits. Assume that this variable has a Poisson distribution. What is the model fit?

(b) State the fitted model and include all estimated parameters. What random-effects parameters are significant? What fixed-effects parameters are significant? Use $\alpha = 0.05$.

(c) Give interpretation of the estimated significant regression coefficients. Did the intervention work?

(d) Find the predicted number of consumed fruits for a female student in the intervention group on the 30th day.

EXERCISE 9.5. A new general manager of a hotel chain is interested in measuring and improving hotel performance. He obtains a year worth of data on 16 hotels in the same tier, and records seasonal values for such hotel performance metrics as the average daily rate (ADR) defined as the average rental income per paid occupied room (in U.S. dollars), and the number of days the hotel occupancy rate (OCR) was below 65%. He also notes if the hotels are situated in rural or urban areas. The data for the four seasons (summer, fall, winter, and spring) are:

Hotel	Area	ADR1	OCR1	ADR2	OCR2	ADR3	OCR3	ADR4	OCR4
1	Rural	88	3	76	8	74	11	78	17
2	Rural	79	5	98	9	72	7	54	14
3	Rural	84	2	67	4	64	9	98	13
4	Rural	79	3	88	4	77	3	66	15
5	Rural	68	1	75	8	58	14	80	21
6	Rural	82	0	95	4	85	9	90	16
7	Rural	92	4	93	8	87	13	92	20
8	Rural	58	0	54	9	67	19	84	25
9	Rural	84	1	87	9	94	6	92	19
10	Urban	112	1	137	11	114	5	137	23
11	Urban	104	1	176	8	97	6	146	18
12	Urban	95	3	131	5	135	6	137	11
13	Urban	128	1	113	10	125	3	126	9
14	Urban	96	2	152	10	145	5	153	10
15	Urban	98	0	170	9	129	3	148	16
16	Urban	119	2	121	8	128	6	147	18

(a) Fit a random slope and intercept model (or random intercept-only model, if appropriate) for the days with occupancy below 65%. Use the negative binomial distribution. Discuss the model fit.

(b) State the fitted model. Identify all significant parameters. Use $\alpha = 0.05$. Are responses over seasons correlated within each hotel?

(c) Interpret the estimates of all significant beta coefficients.

(d) Predict the number of days with occupancy rate below 65% for the winter season in a rural hotel with average daily rate of $75.

EXERCISE 9.6. As part of her dissertation, a Ph.D. student in the School of Pharmacy set up a study to evaluate effectiveness of medication adherence educational classes. Twenty eight patients diagnosed with diabetes took part in the study. They were referred to the study by their doctors because their medication adherence for the initial refill was relatively poor. The student conducted an educational class, and throughout the study collected data on participants' gender, age, level of education, and proportion of days that they took diabetes medication for 60-day refills (called proportion days covered or PDC). Some participants had only two refills, some had three, and some had four. The data are:

ID	Gender	Age	Educ	PDC1	PDC2	PDC3	PDC4
1	F	46	HS+	0.28	0.38	0.55	0.93
2	M	37	<HS	0.12	0.30	0.42	0.80
3	M	45	HS+	0.37	0.62	0.75	0.92
4	F	47	HS+	0.37	0.52	0.52	.
5	F	42	HS+	0.38	0.52	0.38	.
6	M	41	HS+	0.03	0.73	0.27	0.80
7	M	37	HS+	0.27	0.42	0.50	0.77
8	F	45	<HS	0.10	0.52	0.28	.
9	M	49	<HS	0.07	0.27	0.45	.
10	M	49	<HS	0.25	0.30	0.27	0.85
11	M	36	<HS	0.05	0.43	0.35	0.67
12	M	49	>HS	0.03	0.57	.	.
13	M	43	>HS	0.23	0.75	0.87	0.90
14	F	40	>HS	0.07	0.48	0.27	0.62
15	F	39	HS+	0.38	0.53	0.28	0.58
16	M	41	HS+	0.33	0.30	0.45	0.68
17	M	45	<HS	0.27	0.33	.	.
18	F	49	>HS	0.25	0.72	0.28	0.78
19	F	43	HS+	0.37	0.38	0.55	0.68
20	M	39	<HS	0.05	0.30	0.30	0.62
21	F	49	>HS	0.37	0.63	0.28	0.88
22	F	36	HS+	0.08	0.72	0.87	0.98
23	M	40	>HS	0.07	0.73	.	.
24	F	38	HS+	0.35	0.38	0.25	0.63
25	M	36	HS+	0.18	0.52	0.27	0.83
26	F	39	>HS	0.30	0.65	.	.
27	F	40	>HS	0.40	0.60	.	.
28	F	47	HS+	0.22	0.38	0.43	0.73

(a) Run the random slope and intercept (possibly random intercept-only) model for the response variable PDC, using a beta distribution. Does the model fit the data well?

(b) Specify the fitted model and interpret all estimated significant fixed-effects parameters. Use the significance level of 0.05.

(c) What is the predicted PDC value for the second refill of medication for a 50-year-old man with a Bachelor's degree?

EXERCISE 9.7. Use the data in Exercise 9.1 to do the following:

(a) Fit the GEE models with the gamma underlying distribution and with unstructured, Toeplitz, autoregressive, exchangeable, and independent working correlation structures for the response variable EWL.

(b) Compare model fits. Use the QIC criterion.
(c) For the model that has the best fit, do questions (c)-(e) from Exercise 9.1.

EXERCISE 9.8. For the data from Exercise 9.2,
(a) Run the Generalized Estimating Equations models based on a gamma distribution of the response variable Exercise. Try the unstructured, Toeplitz, autoregressive, exchangeable, and independent working correlation matrices.
(b) Determine the model with the best fit with respect to the QIC criterion.
(c) Redo parts (c)-(e) of Exercise 9.2 for the best-fitted model.

EXERCISE 9.9. Use the data in Exercise 9.3 to carry out the following analysis:
(a) Fit the Generalized Estimating Equations models for logistically distributed presence or absence of side effects, with unstructured, Toeplitz, autoregressive, exchangeable, and independent working correlation matrices.
(b) Choose the best model according to the QIC value.
(c) For the best-fitted model, answer questions (b)-(d) from Exercise 9.3.

EXERCISE 9.10. Answer the questions below using the data from Exercise 9.4.
(a) Fit the GEE models based on a Poisson distribution. Use Fruits as the response variable, and try to fit unstructured, Toeplitz, autoregressive, exchangeable, and independent working correlation structures for this response.
(b) Which of the fitted models has the smallest QIC value?
(c) For the best-fitted model, do parts (b)-(d) in Exercise 9.4.

EXERCISE 9.11. Consider the data given in Exercise 9.5. Answer the questions below.
(a) Fit a Generalized Estimating Equations model for the days with occupancy below 65% based on a negative binomial distribution. Try different working correlation matrices: unstructured, Toeplitz, autoregressive, exchangeable, and independent.
(b) Choose the QIC-optimal model.
(c) Answer parts (b)-(d) in Exercise 9.5 for the optimal model.

Chapter 10

Hierarchical Regression Model

In this chapter, we introduce models in which observations may be collected once or repeatedly for each individual, but individuals are logically aggregated in some way, possibly at multiple levels (for example, students are clustered within classrooms, which are clustered within schools). These models incorporate potential correlation among observations at each hierarchical level.

10.1 Hierarchical Regression Model for Normal Response

10.1.1 Model Definition

We assume that data are collected at three levels, and, for ease of presentation, will formulate a three-level hierarchical regression model for a special case when data are collected longitudinally for individuals who are clustered in some way. Suppose data are recorded at times $t_1, \ldots, t_p$ on each of n individuals, who are grouped into c clusters. For each individual, there are k predictor variables and one normally distributed response. An observation at time j for individual i in cluster m is $(x_{1ijm}, \ldots, x_{kijm}, y_{ijm})$, where $i = 1, \ldots, n$, $j = 1, \ldots, p$, and $m = 1, \ldots, c$. Some of the x variables may be characteristics of times (level 1), or of individuals (level 2), or of clusters (level 3). A general form of the *three-level hierarchical model* (also termed *three-stage hierarchical regression model* or *multilevel model* or *model for clustered data*) for a normal response is:

$$y_{ijm} = \beta_0 + \beta_1 \, x_{1ijm} + \cdots + \beta_k \, x_{kijm} + \beta_{k+1} \, t_j + u_{1im} + u_{2im} \, t_j + \tau_{1m} + \tau_{2m} \, t_j + \varepsilon_{ijm} \tag{10.1}$$

where $u_{1im} \overset{iid}{\sim} \mathcal{N}(0, \sigma_{u_1}^2)$, $u_{2im} \overset{iid}{\sim} \mathcal{N}(0, \sigma_{u_2}^2)$, $\mathbb{C}ov(u_{1im}, u_{2im}) = \sigma_{u_1 u_2}$, and $\mathbb{C}ov(u_{1im}, u_{2i'm'}) = 0$ for $i \neq i'$. These two random variables are the *level-2 random slope and intercept*, respectively. Also, $\tau_{1m} \overset{iid}{\sim} \mathcal{N}(0, \sigma_{\tau_1}^2)$, $\tau_{2m} \overset{iid}{\sim} \mathcal{N}(0, \sigma_{\tau_2}^2)$,

$\mathbb{C}ov(\tau_{1m}, \tau_{2m}) = \sigma_{\tau_1 \tau_2}$, and $\mathbb{C}ov(\tau_{1m}, \tau_{1m'}) = 0$ for $m \neq m'$. These variables are, respectively, the *level-3 random slope and intercept*. The random errors ε_{ijm}'s are independent with $\mathcal{N}(0, \sigma^2)$ distribution. In addition, all u's are independent of τ's, and both are independent of ε's.

As defined in the above formula, the index i for individuals (or, more generally, level-2 variable) ranges between 1 and n. It is also possible to enumerate individuals only within each cluster: $i = 1, \ldots, n_m$ where $\sum_{m=1}^{c} n_m = n$.

10.1.2   Fitted Model, Interpretation of Estimated Regression Coefficients, Model Goodness-of-Fit Check, Predicted Response

A fitted hierarchical model has the estimated mean response of the form $\widehat{\mathbb{E}}(y) = \widehat{\beta}_0 + \widehat{\beta}_1 x_1 + \cdots + \widehat{\beta}_k x_k + \widehat{\beta}_{k+1} t$. Estimated regression coefficients are interpreted identically to how they are interpreted in a general linear regression model (see Section 1.5). The log-likelihood deviance test is used to check the model fit. The null model in this case has only fixed-effects terms. As expected, the predicted response for a given set of predictors $x_1^0, \ldots, x_k^0$, and t^0 can be calculated as $y^0 = \widehat{\beta}_0 + \widehat{\beta}_1 x_1^0 + \cdots + \widehat{\beta}_k x_k^0 + \widehat{\beta}_{k+1} t^0$.

10.1.3   SAS Implementation

The procedure `mixed` fits hierarchical regression with a normally distributed response. To accommodate the hierarchical structure, `random` statements should be included for variables defining levels 2 and 3 of the model. The full syntax is presented below.

```
proc mixed data=longform_data_name covtest;
    class cluster_name individual_name <list of categorical predictors>;
        model response_name=<list of predictors> time_name/solution
                            outpm=outdata_name;
random intercept time_name/subject=cluster_name type=un;
random intercept time_name/subject=individual_name(cluster_name)
                            type=un;
run;
```

- For prediction, the long-form data set must contain a row with specified values of all the x variables and the time variable. And in addition, one has to specify some (sometimes hypothesized) values for a cluster and an individual within this cluster, possibly defining a new cluster with individual number 1.

10.1.4 R Implementation

Hierarchical model for normally distributed response variable may be fitted to a long-form data set with function `lmer()` in the library `lme4`. The syntax is:

summary(*fitted.model.name*<- lmer(*response.name* ~ *x1.name* + ⋯
+ *xk.name* + *time.name* + (1 + *time.name*|*cluster.name*)
+ (1 + *time.name*|*cluster.name*:*individual.name*), data = *data.name*)).

- As usually, prediction can be carried out using the function `predict()`. In the `data.frame` argument, however, some values for a cluster and an individual must be specified. Also, since a new cluster and individual are added, the argument `allow.new.levels=TRUE` must be included. The syntax looks like this:

predict(*fitted.model.name*, data.frame(*cluster.name*=*new.value*,
individual.name=1, *catpredictor1.name*="*value*", ...,
numpredictork.name=*value*, *time.name*=*value*), allow.new.levels=TRUE)

10.1.5 Example

EXAMPLE 10.1. Mothers and daughters from 24 families with signs of depression were invited for a study of efficacy of a new method of intensive psychotherapy. At the baseline, one- and three-month visits, the quality of life (QOL) questionnaire was filled out by each of the participant, and a QOL score was computed. Higher values of this score indicate better quality of life. Whether signs of depression were present was also recorded (1=present, or 0=absent). The study was done on mother-daughter dyads. This type of study is called *familial* or *dyadic*. The scores are logically expected to be correlated over time for each individual, and also members of the same family might have correlated responses. For some families, more than one daughter participated. Below we develop a three-stage hierarchical model for the data.

```
data dyads;
input family individual relation$ depression1
depression2 depression3 qol1 qol2 qol3 @@;
cards;
1   1 M 1 1 1 4.0 4.1 4.9   1   2 D 1 1 0 2.5 3.2 4.2
2   1 M 1 1 1 2.6 2.8 4.1   2   2 D 1 1 1 2.8 3.1 4.2
3   1 M 1 1 1 2.5 3.8 4.0   3   2 D 1 1 1 2.4 5.1 3.3
4   1 M 1 0 0 2.1 3.3 4.6   4   2 D 1 0 0 3.7 3.1 4.4
5   1 M 1 0 0 2.9 4.2 3.4   5   2 D 1 0 0 2.4 2.6 2.7
6   1 M 1 1 0 3.3 4.2 4.7   6   2 D 1 1 0 2.7 4.0 4.1
7   1 M 1 1 0 3.7 4.3 3.8   7   2 D 1 1 0 2.8 3.2 3.6
8   1 M 1 0 0 3.5 4.1 4.3   8   2 D 1 1 0 1.6 2.6 3.5
9   1 M 1 0 0 4.0 4.4 3.6   9   2 D 1 0 1 1.8 2.5 3.1
10  1 M 1 1 1 3.0 3.7 4.3   10  2 D 1 1 1 2.2 2.0 3.3
11  1 M 1 1 1 4.3 5.0 3.7   11  2 D 1 1 1 3.3 2.5 3.2
12  1 M 1 1 1 3.5 5.4 4.7   12  2 D 1 1 1 3.5 3.6 4.2
13  1 M 1 0 0 4.1 4.5 3.2   13  2 D 1 0 0 3.7 4.2 3.5
14  1 M 1 1 0 5.0 4.2 3.6   14  2 D 1 0 0 3.3 4.6 3.0
15  1 M 1 1 0 1.8 2.2 2.3   15  2 D 1 0 0 2.4 3.5 3.6
16  1 M 1 0 0 3.1 2.5 3.9   16  2 D 1 1 0 2.0 2.9 2.4
17  1 M 1 1 0 3.4 5.5 4.7   17  2 D 1 0 1 3.2 4.3 3.7
18  1 M 1 0 0 3.4 5.3 4.1   18  2 D 1 1 0 1.8 3.4 3.1
19  1 M 1 0 0 3.5 3.3 5.1   19  2 D 1 0 0 2.8 4.3 3.4
20  1 M 1 0 0 3.5 3.3 5.1   20  2 D 1 0 0 3.2 4.9 3.6
21  1 M 1 0 0 2.9 2.7 3.5   21  2 D 1 0 0 4.3 3.7 2.5
22  1 M 1 0 0 4.8 4.3 4.3   22  2 D 1 0 0 4.5 3.8 3.3
23  1 M 1 1 0 3.6 3.9 3.7   23  2 D 1 0 0 3.7 3.7 3.5
23  3 D 1 1 0 2.5 1.8 2.3   24  1 M 1 1 0 5.0 4.4 4.2
24  2 D 1 1 0 4.9 3.2 2.5   24  3 D 1 0 0 2.7 3.1 4.1
24  4 D 1 0 1 3.0 3.5 3.6
;

data longform;
 set dyads;
array d[3] depression1-depression3;
 array q[3] qol1-qol3;
  do visit=1 to 3;
    depression=d[visit];
     qol_score=q[visit];
   output;
  end;
keep family individual relation depression visit qol_score;
```

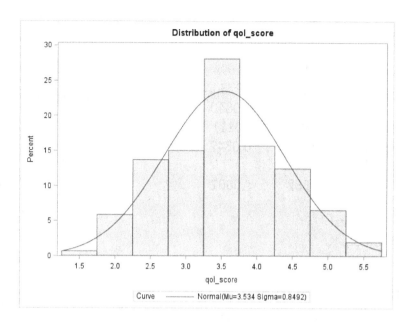

Figure 10.1: Histogram for QOL_score in SAS

```
run;

proc univariate;
 var qol_score;
  histogram/normal;
run;
```

Goodness-of-Fit Tests for Normal Distribution
Test	p Value
Kolmogorov-Smirnov	0.150
Cramer-von Mises	>0.250
Anderson-Darling	>0.250

The tests support normality, and the histogram shows a roughly bell-shaped curve. Therefore, we will fit a hierarchical model based on normal distribution.

```
proc mixed covtest;
 class family individual relation(ref='D') depression;
  model qol_score=relation depression visit/solution;
 random intercept visit/subject=family type=un;
random intercept visit/subject=individual(family) type=un;
run;
```

Covariance Parameter Estimates

Cov Parm	Subject	Estimate	Pr Z
UN(1,1)	family	0.2827	0.1694
UN(2,1)	family	-0.1238	0.2730
UN(2,2)	family	0.06723	0.0840
UN(1,1)	individual(family)	0.4795	0.0111
UN(2,1)	individual(family)	-0.09011	0.0342
UN(2,2)	individual(family)	0	.
Residual		0.3882	<.0001

Null Model Likelihood Ratio Test

DF	Chi-Square	Pr > ChiSq
5	17.83	0.0032

Solution for Fixed Effects

Effect	mother	depression	Estimate	Pr > |t|
Intercept			2.7180	<.0001
relation	M		0.6060	<.0001
relation	D		0	.
depression		0	0.04226	0.7777
depression		1	0	.
visit			0.2569	0.0150

This model has a reasonable fit since the p-value in the likelihood ratio test is smaller than 0.05. At the individual level, the variance of the random intercept, and covariance between the intercept and the slope are significant at the 5% level. At the family level, variance of the random slope is significant at the 10% level. Thus, at the 5% significance level, observations are correlated for each individual, but not between family members. At the 10% level, however, correlation between family members also exists. The estimated mean QOL score may be written as $\widehat{\mathbb{E}}(QOL) = 2.7180 + 0.6060 * mother + 0.04226 * no\_depression + 0.2569 * visit$. At the 5% level, depression is not a significant predictor of quality of life, whereas both mother and visit are significant predictors. Estimated average QOL score for mothers is larger than that for daughters by 0.6060, and the estimated score increases, on average, by 0.2569 between visits.

Suppose now that we would like to predict the QOL score for a mother at visit 3 who doesn't show any symptoms of depression. We calculate $QOL^0 = 2.7180 + 0.6060 + 0.04226 + 0.2569 * 3 = 4.13696$. SAS outputs similar value as seen below.

```
data prediction;
input family individual relation$ depression visit;
cards;
25 1 M 0 3
run;

data longform;
 set longform prediction;
run;

proc mixed covtest;
 class family individual relation(ref='D') depression;
  model qol_score=relation depression visit/solution
  outpm=outdata;
 random intercept visit/subject=family type=un;
 random intercept visit/subject=individual(family) type=un;
run;

proc print data=outdata (firstobs=154 obs=154);
var Pred;
run;

    Pred
4.13688
```

The R script and output for fitting the hierarchical model are:

```
dyads.data<-read.csv(file="./Example10.1Data.csv",
header=TRUE, sep=",")

#creating long-form data set
install.packages("reshape2")
library(reshape2)

data.depr<- melt(dyads.data[,c("family", "individual",
 "relation","depression1","depression2", "depression3")],
id.vars=c("family", "individual", "relation"),
variable.name="depr.visits", value.name="depression")
data.qol<- melt(dyads.data[,c("qol1","qol2", "qol3")],
variable.name="qol.visits", value.name="qol")
longform.data<- cbind(data.depr, data.qol)
```

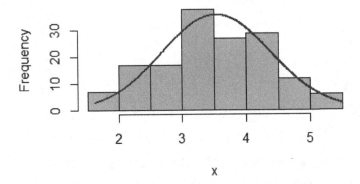

Figure 10.2: Histogram for QOL in R

```
#creating numeric variable for time
visit<- ifelse(longform.data$depr.visits=="depression1", 1,
ifelse(longform.data$depr.visits=="depression2", 2, 3))

#plotting histogram with fitted normal density
install.packages("rcompanion")
library(rcompanion)

plotNormalHistogram(longform.data$qol)

#testing for normality of distribution
shapiro.test(longform.data$qol)
```

Shapiro-Wilk normality test
W = 0.98955, p-value = 0.3147

```
#specifying reference category
depression.rel<- relevel(as.factor(longform.data$depression),
ref="1"))

#fitting hierarchical model
install.packages("lme4")
library(lme4)

summary(fitted.model<- lmer(qol ~ relation + depression.rel
+ visit + (1 + visit|family)+ (1 + visit|family:individual),
data=longform.data))
```

Random effects:

Groups	Name	Variance	Corr
family:individual	(Intercept)	0.54480	
	visit	0.03518	-1.00
family	(Intercept)	0.28400	
	visit	0.05722	-0.90
Residual		0.35886	

Fixed effects:

	Estimate	t value
(Intercept)	2.71793	12.731
relationM	0.60277	4.458
depression.rel0	0.03476	0.235
visit	0.25838	2.651

```
#checking model fit
null.model<- glm(qol ~ relation + depression.rel + visit,
data=longform.data)
print(deviance<- -2*(logLik(null.model)-logLik(fitted.model)))
```

5.332274

```
print(p.value<- pchisq(deviance, df=5, lower.tail=FALSE))
```

0.3766883

```
#using fitted model for prediction
print(predict(fitted.model, data.frame(family=25, individual=1,
relation="M", depression.rel="0", visit=3),
allow.new.levels=TRUE))
```

4.130604

□

10.2 Hierarchical Regression Model for Non-normal Response

10.2.1 Model Definition

A hierarchical regression model for the response variable y_{ijm} that has a gamma, or logistic, or Poisson, or negative binomial, or beta distribution is defined similarly to a normally distributed response but the expression (10.1) takes the form

$$g\big(\mathbb{E}(y_{ijm})\big) = \beta_0 + \beta_1\,x_{1ijm} + \cdots + \beta_k\,x_{kijm} + \beta_{k+1}\,t_j + u_{1im} + u_{2im}\,t_j + \tau_{1m} + \tau_{2m}\,t_j$$

where $g(\cdot)$ is the link function that corresponds to the underlying distribution.

10.2.2 Fitted Model

A fitted hierarchical model with the link function $g(\cdot)$ is

$$g\big(\widehat{\mathbb{E}}(y)\big) = \widehat{\beta}_0 + \widehat{\beta}_1\,x_1 + \cdots + \widehat{\beta}_k\,x_k + \widehat{\beta}_{k+1}\,t. \tag{10.2}$$

10.2.3 Interpretation of Estimated Regression Coefficients

In view of (10.2), in the hierarchical regression model with non-normal response, estimated regression coefficients are interpreted identically to how they are interpreted in a generalized linear regression model, in terms of a change or difference in the link function of the estimated mean response. Or, more specifically, the same way as estimated regression coefficients are interpreted for each of the appropriate regressions models: gamma (Subsection 2.2.3), logistic (Subsection 3.1.3), Poisson (Subsection 5.1.3), negative binomial (Subsection 6.1.3), and beta (Subsection 7.1.3).

10.2.4 Model Goodness-of-Fit Check

The model fit can be verified via the standard log-likelihood deviance test, where the null model has no random-effects terms.

10.2.5 Predicted Response

By (10.2), the predicted response for a given set of predictors $x_1^0, \ldots, x_k^0$, and t^0 is found as $y^0 = g^{-1}\big(\widehat{\beta}_0 + \widehat{\beta}_1\,x_1^0 + \cdots + \widehat{\beta}_k\,x_k^0 + \widehat{\beta}_{k+1}\,t^0\big)$ for the appropriate link function $g(\cdot)$.

10.2.6 SAS Implementation

The syntax for procedure `glimmix` presented in Subsection 9.1.3 applies directly to fitting hierarchical regression models. Two **random** statements should be included for variables defining levels 2 and 3 of the model. The general syntax is:

```
proc glimmix data=data_name method=Laplace;
class cluster_name individual_name catpredictor1_name
(ref='level_name') catpredictor2_name (ref='level_name') ...;
model response_name=<list    of      predictors> time_name/solution
dist=dist_name link=link_type;
    output out=outdata pred(ilink)=predicted_name;
random intercept time_name/subject=cluster_name type=un;
random intercept time_name/subject=individual_name(cluster_name)
type=un;    covtest/wald;
run;
```

- For prediction purposes, one has to add to the long-form data set a row containing a value for cluster, a value for individual, and values for all predictor and time variables.

10.2.7   R Implementation

Function `glmer()` in the library `lme4` fits a hierarchical model for a response variable with a specified distribution. The statements are:

summary(*fitted.model.name*<- glmer(*response.name* ~ *x1.name* + ⋯
+ *xk.name* + *time.name* + (1 + *time.name*|*cluster.name*)
+ (1 + *time.name*|*cluster.name*/*individual.name*), data=*data.name*,
family=*dist.name*(link="*link.type*")))

- The choices for distribution names and link types are `gamma(link="log")`, `binomial(link="logit")`, `poisson(link="log")`, `nb(link="log")`, and `beta(link="logit")`.

- The syntax for prediction is:

predict(*fitted.model.name*, data.frame(*cluster.name*=*new.value*, *individual.name*=1, *catpredictor1.name*="*value*", ..., *numpredictork.name*=*value*, *time.name*=*value*), re.form=NA, type="response")

10.2.8   Example

EXAMPLE 10.2. For the long-form data set created in the previous example, we fit a hierarchical model based on logistic distribution to model presence of symptoms of depression. First we try a complete three-stage model. The code is given below. This model, however, doesn't converge.

```
proc glimmix method=Laplace;
 class family individual relation(ref='D');
    model depression=relation qol_score visit/solution
   dist=binomial link=logit;
random intercept visit/subject=family type=un;
random intercept visit/subject=individual(family) type=un;
   covtest/wald;
run;
```

The most complex model that converges is the one with the intercept at the family level only.

```
proc glimmix method=Laplace;
 class family individual relation(ref='D');
  model depression=relation qol_score visit/solution
            dist=binomial link=logit;
random intercept/subject=family type=un;
   covtest/wald;
run;
```

-2 Log Likelihood 123.46

Covariance Parameter Estimates

Cov Parm	Subject	Estimate	Pr > Z
UN(1,1)	family	5.9549	0.0491

Solutions for Fixed Effects

Effect	mother	Estimate	Pr > |t|
Intercept		9.3445	0.0009
relation	M	0.2761	0.6400
relation	D	0	.
qol_score		-0.5214	0.2576
visit		-3.2363	<.0001

The variance of the random intercept at the family level is statistically significant at the 10% level. Next, we conduct the goodness-of-fit test.

```
proc glimmix;
 class family individual relation(ref='D');
  model depression=relation qol_score visit/solution
            dist=binomial link=logit;
run;
```

-2 Log Likelihood 143.50

```
data deviance_test;
 deviance=143.50-123.46;
  pvalue=1-probchi(deviance,6);
run;

proc print;
run;
```

deviance   pvalue
 20.04   .002724357

The p-value is less than 0.05, indicating a good fit. The fitted model has the form

$$\widehat{\mathbb{P}}(depr=1) = \frac{\exp\{9.3445 + 0.2761 * mother - 0.5214 * qol\_score - 3.2363 * visit\}}{1 + \exp\{9.3445 + 0.2761*mother - 0.5214*qol\_score - 3.2363 * visit\}}.$$

Significant predictor is only the visit number. The estimated odds in favor of presence of symptoms of depression change by $\left(\exp\{-2.4940\} - 1\right) \cdot 100\% = -91.74\%$, that is, decrease by 91.74% from visit to visit.

The predicted probability that a mother with the quality of life score of 3.5 shows symptoms of depression during the third visit is

$$\mathbb{P}^0(depr) = \frac{\exp\{9.3445 + 0.2761 - 0.5214 * 3.5 - 3.2363 * 3\}}{1 + \exp\{9.3445 + 0.2761 - 0.5214 * 3.5 - 3.2363 * 3\}} = 0.12862.$$

The prediction in SAS is carried out via the following statements.

```
data prediction;
input family individual relation$ qol_score visit;
cards;
25 1 M 3.5 3
;

data longform;
 set longform prediction;
run;

proc glimmix method=Laplace;
 class family individual relation(ref='D');
  model depression=relation qol_score visit/solution
            dist=binomial link=logit;
```

```
random intercept/subject=family type=un;
   covtest/wald;
 output out=outdata pred(ilink)=p_probdepr;
run;

proc print data=outdata (firstobs=154 obs=154);
var p_probdepr;
run;
```

p_probdepr
   0.12862

R script and relevant output for this example follow.

```
dyads.data<- read.csv(file="./Example10.1Data.csv",
header=TRUE, sep=",")

#creating long-form data set
install.packages("reshape2")
library(reshape2)

data.depr<- melt(dyads.data[,c("family", "individual",
"relation",
"depression1", "depression2", "depression3")],
 id.vars=c("family",
"individual", "relation"), variable.name="depr.visits",
value.name="depression")
data.qol<- melt(dyads.data[,c("qol1","qol2", "qol3")],
variable.name="qol.visits", value.name="qol")
longform.data<- cbind(data.depr, data.qol)

#creating numeric variable for time
visit<- ifelse(longform.data$depr.visits=="depression1", 1,
ifelse(longform.data$depr.visits=="depression2", 2, 3))

#fitting hierarchical model (fails to converge)
install.packages("lme4")
library(lme4)

summary(glmer(depression ~ relation + qol + visit
+ (1 + visit|family)
+ (1 + visit|family/individual), data=longform.data,
```

```
family=binomial(link="logit")))

#fitting simpler hierarchical model (converges)
summary(fitted.model<- glmer(depression ~ relation
+ qol + visit + (1|family), data=longform.data,
family=binomial(link="logit")))
```

Random effects:

Groups	Name	Variance
family	(Intercept)	5.955

Fixed effects:

	Estimate	Pr(>|z|)
(Intercept)	9.3444	0.000138
mother	0.2761	0.639248
qol	-0.5214	0.255404
visit	-3.2363	2.24e-07

```
#checking model fit
null.model<- glm(depression ~ relation + qol + visit,
data=longform.data, family=binomial(link=logit))
print(deviance<- -2*(logLik(null.model)-logLik(fitted.model)))
```

20.03607

```
print(p.value<- pchisq(deviance, df=6, lower.tail=FALSE))
```

0.00272875

```
#using fitted model for prediction
print(predict(fitted.model, data.frame(family=25, individual=1,
relation="M", qol=3.5, visit=3), re.form=NA, type="response"))
```

0.1286211

□

Exercises for Chapter 10

EXERCISE 10.1. For the hierarchical model with normal response defined in (10.1), show that
(a) The observations within each individual i in cluster m for different times j and j' have covariance $\mathbb{C}\mathrm{ov}(y_{ijm}, y_{ij'm}) = \sigma_{u_1}^2 + \sigma_{u_1 u_2}(t_j + t_{j'}) + \sigma_{u_2}^2 + \sigma_{\tau_1}^2 + \sigma_{\tau_1 \tau_2}(t_j + t_{j'}) + \sigma_{\tau_2}^2 t_j t_{j'}$.

(b) The observations for two individuals i and i' within the same cluster m at any two times t_j and $t_{j'}$, equal or not, have covariance $\mathbb{C}ov(y_{ijm}, y_{i'j'm}) = \sigma_{\tau_1}^2 + \sigma_{\tau_1\tau_2}(t_j + t_{j'}) + \sigma_{\tau_2}^2 t_j t_{j'}$.

(c) Observations for two individuals in different clusters are not correlated, that is, $\mathbb{C}ov(y_{ijm}, y_{i'j'm'}) = 0$ where $i \neq i'$ and $m \neq m'$.

(d) The response variable y_{ijm} has a normal distribution with the mean $\mathbb{E}(y_{ijm}) = \beta_0 + \beta_1 x_{1ijm} + \cdots + \beta_k x_{kijm} + \beta_{k+1} t_j$ and variance $\mathbb{V}ar(y_{ijm}) = \sigma_{u_1}^2 + 2\sigma_{u_1u_2} t_j + \sigma_{u_2}^2 + \sigma_{\tau_1}^2 + 2\sigma_{\tau_1\tau_2} t_j + \sigma_{\tau_2}^2 t_j^2 + \sigma^2$.

EXERCISE 10.2. A team of school inspectors is studying scores on tests in English Language Arts (ELA), Mathematics, and Science and their relation to schools' Academic Performance Index (API) and classroom size. Data on average classroom scores for five consecutive years at two schools are available.

School	API	Subject	ClassSize	Year	Score
1	911	ELA	20	13	78.39
1	912	ELA	22	14	79.85
1	917	ELA	23	15	81.34
1	917	ELA	22	16	82.56
1	919	ELA	24	17	83.12
1	911	Math	21	13	83.77
1	912	Math	22	14	84.9
1	917	Math	24	15	86.12
1	917	Math	23	16	88.99
1	919	Math	23	17	88.4
1	911	Science	21	13	80.19
1	912	Science	22	14	83.15
1	917	Science	24	15	84.45
1	917	Science	23	16	86.66
1	919	Science	23	17	88.43
2	732	ELA	34	13	68.03
2	745	ELA	36	14	70.67
2	751	ELA	36	15	74.17
2	753	ELA	37	16	72.78
2	753	ELA	38	17	73.18
2	732	Math	34	13	67.88
2	745	Math	34	14	68.34
2	751	Math	35	15	70.3
2	753	Math	37	16	71.22
2	753	Math	36	17	72.12
2	732	Science	34	13	72.96
2	745	Science	34	14	73.65
2	751	Science	36	15	74.58
2	753	Science	35	16	76.36
2	753	Science	35	17	76.23

(a) Plot a histogram for test scores and conduct normality testing. Verify that the underlying distribution may be modeled as normal.

(b) Run the model with random intercepts at the school and subject-within-school levels. Check first that models with random slopes do not converge. Use $\alpha = 0.10$ to draw conclusion about significant parameters. Are the scores for each subject correlated? Are the scores within each school correlated?

(c) Write down the fitted model and interpret all estimated significant fixed-effects coefficients.

(d) Use the fitted model to predict average score on a math test for a class of 36 students in 2017 in a school with an API of 753.

EXERCISE 10.3. An environmental health scientist studies levels of pollution. For the study he randomly chooses three states and two counties within each state. In each county he randomly samples eight townships, and measures levels of an atmospheric particulate matter that have a diameter of less than 2.5 micrometers (PM2.5). For each township, he also records population size (in thousands of people) and whether pesticides are applied on a large scale in farming. The collected data are summarized in the following table.

State	County	Twnshp	Popl	Pest	PM2.5	State	County	Twnshp	Popl	Pest	PM2.5
S1	A	1	4.1	no	2.97	S2	D	1	6.6	no	9.13
S1	A	2	2.0	no	3.05	S2	D	2	7.2	no	11.04
S1	A	3	6.3	no	4.97	S2	D	3	8.3	no	8.98
S1	A	4	3.2	no	3.77	S2	D	4	5.2	yes	5.75
S1	A	5	3.4	no	3.09	S2	D	5	9.1	yes	11.28
S1	A	6	3.9	yes	4.75	S2	D	6	4.3	no	6.88
S1	A	7	3.8	yes	6.93	S2	D	7	6.9	yes	9.21
S1	A	8	5.6	yes	5.83	S2	D	8	8.5	yes	11.23
S1	B	1	12.7	no	13.19	S3	E	1	6.1	no	5.44
S1	B	2	17.8	no	12.9	S3	E	2	3.9	no	4.33
S1	B	3	13.7	no	11.45	S3	E	3	3.5	no	5.04
S1	B	4	11.8	yes	15.4	S3	E	4	2.4	no	3.31
S1	B	5	12.9	yes	14.15	S3	E	5	4.3	no	5.24
S1	B	6	13	yes	15.16	S3	E	6	2.8	yes	4.34
S1	B	7	12	yes	14.36	S3	E	7	3.4	no	3.9
S1	B	8	13	no	14.38	S3	E	8	3.6	no	3.59
S2	C	1	9.9	no	7.25	S3	F	1	5.3	no	5.01
S2	C	2	5.6	yes	8.46	S3	F	2	4.5	no	5.73
S2	C	3	3.9	no	7.06	S3	F	3	2.5	no	4.28
S2	C	4	7.3	no	9.33	S3	F	4	3.1	yes	5.42
S2	C	5	4.7	no	5.59	S3	F	5	3.5	no	3.59
S2	C	6	8.9	yes	9.94	S3	F	6	5.7	no	4.69
S2	C	7	6.7	yes	8.49	S3	F	7	7.1	no	4.06
S2	C	8	6.5	yes	6.97	S3	F	8	4.6	no	3.98

(a) Plot a histogram of the response variable PM2.5. Describe its shape. Argue that gamma distribution is appropriate.

(b) Run the multilevel regression model for PM2.5, based on the gamma distribution. How many levels are there? How well does the model fit the data? Hint: Townships variable indexes repeated measures within each county.

(c) Write down the fitted model. Specify all estimates. Are PM2.5 readings correlated within each county? Within each state?

(d) What fixed-effects predictors are significant? Interpret them.

(e) Use the fitted model to predict the level of particulate matter in a town with population of 2,500 people that is located in County A, if it is known that no pesticides are used in the fields that surround this town.

EXERCISE 10.4. A financial analyst is studying behavior of certain portfolios with stocks, bonds, and currency. He records whether the prices of the assets went up at the closure of the stock exchange at the end of five consecutive business days (1=went up, 0=went down or stayed the same). The data follow.

Portfolio	Asset	Type	Day1	Day2	Day3	Day4	Day5
1	1	Stock	0	0	0	0	1
1	2	Stock	0	0	0	0	0
1	3	Bond	0	0	0	1	1
1	4	Bond	0	1	1	1	1
1	5	Stock	0	0	1	1	1
1	6	Stock	0	0	1	1	1
1	7	Stock	0	1	1	1	1
2	8	Currency	1	1	0	0	0
2	9	Stock	0	0	0	1	1
2	10	Bond	1	1	0	1	1
2	11	Stock	1	0	1	1	1
3	12	Currency	1	0	1	0	1
3	13	Stock	0	1	0	0	0
3	14	Stock	0	1	0	0	0
4	15	Stock	1	0	1	0	0
4	16	Bond	1	1	1	1	0
4	17	Currency	0	0	1	1	1
4	18	Stock	0	1	0	0	0
4	19	Currency	0	0	0	1	1
5	20	Stock	1	1	0	1	1
5	21	Currency	1	1	1	1	1
5	22	Stock	1	0	0	1	1
5	23	Bond	1	1	0	1	1
5	24	Stock	0	0	1	1	1

(a) Run a three-level hierarchical model for the binary response variable. Are the measurements correlated within each asset in each portfolio? Within each portfolio? How good is the model fit?

(b) What predictors are significant at the 5% level? Interpret the estimated significant regression coefficients.

(c) According to the fitted model, what are the estimated odds in favor of increase in value on the third day of a currency from the fourth portfolio?

EXERCISE 10.5. A team of child psychologists conducted a study with third graders. They administered a test to high performing students at all third-grade classes at three schools. The computer-based test consisted of four tasks. The researchers were interested in how many additional attempts beyond one it would take each student to complete each task. The measurements are summarized in the following table.

School	Class	Student	Gender	Task1	Task2	Task3	Task4
1	1	1	boy	1	3	4	5
1	1	2	boy	0	0	3	4
1	1	3	boy	1	2	4	5
1	1	4	girl	3	3	5	5
1	1	5	boy	1	1	4	3
1	1	6	girl	2	4	3	4
1	1	7	girl	1	2	3	7
1	2	1	boy	1	2	3	5
1	2	2	boy	0	1	7	6
1	2	3	girl	3	4	8	7
1	2	4	boy	2	2	5	6
1	2	5	girl	3	3	5	8
2	1	1	boy	1	3	4	8
2	1	2	boy	0	5	5	3
2	1	3	girl	2	6	7	9
2	1	4	boy	0	2	4	6
2	1	5	boy	2	3	3	5
2	1	6	girl	3	5	4	9
2	1	7	boy	1	3	7	3
2	2	1	girl	0	2	6	5
2	2	2	boy	0	0	4	3
2	2	3	girl	3	4	7	6
2	2	4	boy	1	2	5	3
2	2	5	boy	3	4	2	2
2	2	6	girl	1	1	6	8
2	2	7	girl	4	3	8	7
2	2	8	girl	2	5	4	5
2	3	1	girl	0	0	2	1
2	3	2	girl	0	1	2	4
2	3	3	boy	0	1	1	3
2	3	4	boy	0	1	0	2
2	3	5	girl	1	1	1	2
2	3	6	boy	0	0	0	1

(a) Fit a four-level hierarchical regression to model the number of additional attempts: level 1 are tasks, level 2 are students, level 3 are classrooms, and level 4 are schools. Assume that the underlying distribution is Poisson. What is the fit of this model?

(b) Present the fitted model. What can you say about the correlation of the repeated measures for each student? Among the students in each classroom? Among the students in each school? Interpret estimated significant fixed-effects coefficients.

(c) Use the fitted model to predict the number of extra attempts it would take a girl in School 2, in Class 3 to complete Task 4.

EXERCISE 10.6. An undergraduate advisor in Mathematics at a Master's degree granting university is interested in finding out how many students who graduate with a Bachelor's degree stay on at the same department to get a Master's degree. She records the data for her university and also sends out a brief survey to Biology, Chemistry, and Mathematics departments at seven comparable universities. The data she obtains are for the most recent three years.

Univ	Dept	Year1	Year2	Year3	Univ	Dept	Year1	Year2	Year3
1	bio	6	13	17	5	bio	7	16	15
1	chem	8	7	12	5	chem	4	4	4
1	math	10	14	13	5	math	8	1	6
2	bio	2	8	8	6	bio	5	3	3
2	chem	0	9	9	6	chem	5	3	4
2	math	0	5	8	6	math	3	2	5
3	bio	2	8	5	7	bio	7	2	8
3	chem	3	3	5	7	chem	9	12	9
3	math	8	9	16	7	math	7	15	16
4	bio	1	11	12	8	bio	3	6	8
4	chem	1	5	4	8	chem	12	11	10
4	math	5	16	17	8	math	8	4	13

(a) Argue that the data may be modeled as having a negative binomial distribution. What quantities support your argument?
(b) Run a multilevel model. Does the model fit the data well?
(c) Are the observations correlated for each department over time? For the departments within the same university? State the fitted model, specifying all parameter estimates.
(d) Does the response change significantly over the years?
(e) What is the predicted number of students who would stay on for Master's program at the first university, in the math department, in year 3.

EXERCISE 10.7. A multi-center clinical trial in pharmacology studies the response for different medications. Same subjects test four medications with proper washout periods observed in-between. The response to medications is measured on a continuous scale. The data are as follows.

Center	Subject	Gender	Med A	Med B	Med C	Med D
1	101	M	0.32	0.27	0.23	0.9
1	102	M	0.27	0.16	0.35	0.5
1	103	F	0.39	0.44	0.45	0.64
1	104	M	0.14	0.47	0.63	0.76
1	105	F	0.38	0.36	0.4	0.42
1	106	F	0.61	0.53	0.64	0.79
1	107	F	0.55	0.73	0.63	0.41
1	108	M	0.4	0.47	0.46	0.99
1	109	F	0.25	0.4	0.31	0.42
1	110	M	0.34	0.48	0.29	0.63
1	111	M	0.33	0.42	0.43	0.35
1	112	F	0.21	0.39	0.74	0.98
1	113	F	0.39	0.22	0.5	0.58
1	114	M	0.33	0.3	0.26	0.19
1	115	F	0.53	0.49	0.36	0.73
2	201	M	0.31	0.46	0.53	0.81
2	202	F	0.4	0.57	0.28	0.84
2	203	M	0.26	0.42	0.38	0.9
2	204	M	0.33	0.34	0.56	0.75
2	205	F	0.29	0.45	0.57	0.81
2	206	F	0.3	0.42	0.64	0.95
2	207	F	0.34	0.42	0.55	0.77
2	208	M	0.19	0.35	0.42	0.67
2	209	M	0.25	0.44	0.62	0.73
2	210	F	0.21	0.41	0.58	0.75
3	301	F	0.23	0.41	0.5	0.86
3	302	F	0.21	0.35	0.52	0.84
3	303	M	0.21	0.43	0.68	0.72
3	304	M	0.27	0.23	0.47	0.59
3	305	M	0.11	0.28	0.5	0.78
3	306	F	0.19	0.24	0.55	0.73
3	307	M	0.15	0.23	0.39	0.82
3	308	F	0.18	0.19	0.53	0.92

(a) Run the multilevel regression to model the response to medication, assuming that it follows a beta distribution.

(b) State the model and estimate the parameters. What random-effects terms are present? Discuss the model fit.

(c) Interpret the results. Are responses correlated for each subject? For each center? Interpret estimated significant fixed-effects terms.

(d) Use the fitted model to predict the response to medication A in a female subject in Center 3.

Recommended Books

[1] Agresti, Alan. *Foundations of Linear and Generalized Linear Models*, Wiley, 2015.

[2] Allison, Paul D. *Logistic Regression Using SAS: Theory and Application*,SAS Institute, 2nd edition, 2012.

[3] Demidenko, Eugene. *Mixed Models: Theory and Applications with R*, Wiley, 2nd edition, 2013.

[4] Dobson, Annette J, and Adrian G. Barnett. *An Introduction to Generalized Linear Models*, Chapman & Hall/CRC Press, 4th edition, 2018.

[5] Faraway, Julian J. *Extending the Linear Model with R: Generalized Linear, Mixed Effects and Nonparametric Regression Models*, Chapman & Hall/CRC Press, 2nd edition, 2016.

[6] Faraway, Julian J. *Linear Models with R*, Chapman & Hall/CRC Press, 2nd edition, 2014.

[7] Finch, W. Holmes, Bolin, Jocelyn E., and Ken Kelley. *Multilevel Modeling Using R*, Chapman & Hall/CRC Press, 2014.

[8] Fox, John. *Applied Regression Analysis and Generalized Linear Models*, SAGE Publications, Inc., 3rd edition, 2015.

[9] Galecki, Andrzej, and Tomasz Burzykowski. *Linear Mixed-Effects Models Using R: A Step-by-Step Approach*, Springer, 2015.

[10] Garson, G. David. *Hierarchical Linear Modeling: Guide and Applications*, SAGE Publications, Inc., 2012.

[11] Hosmer, David W. Jr., Lemeshow, Stanley, and Rodney X. Sturdivant. *Applied Logistic Regression*, Wiley, 3rd edition, 2013.

[12] Lee, Youngjo Lee, Ronnegard, Lars, and Maengseok Noh. *Data Analysis Using Hierarchical Generalized Linear Models with R*, Chapman & Hall/CRC Press, 2017.

[13] Searle, Shayle R. and Marvin H. J. Gruber. *Linear Models*, Wiley, 2016.

[14] Stroup, Walter W. *Generalized Linear Mixed Models: Modern Concepts, Methods and Applications*, Chapman & Hall/CRC Press, 2012.

[15] West, Brady T., Welch, Kathleen B., and Andrzej T. Galecki. *Linear Mixed Models: A Practical Guide Using Statistical Software*, Chapman & Hall/CRC Press, 2nd edition, 2014.

[16] Wood, Simon N. *Generalized Additive Models: An Introduction with R*, Chapman & Hall/CRC, 2nd edition, 2017.

List of Notation

Q, 239
$V(\cdot)$, 237
$\Phi(\cdot)$, 56, 80
$\Phi^{-1}(\cdot)$, 56
α, 37, 238, 239
$\alpha_1, \ldots, \alpha_{c-1}$, 72, 80, 86, 93
β, 37
$\beta_0, \ldots, \beta_k$, 1, 2, 37, 49, 56, 61, 105, 110, 138, 142, 168
$\beta_0, \ldots, \beta_m$, 116, 122, 147, 153, 173, 182, 189
$\beta_0, \ldots, \beta_{k+1}$, 210, 237, 258, 281, 282, 290, 296
$\beta_1, \ldots, \beta_k$, 72, 80, 86
$\beta_{j1}, \ldots, \beta_{jk}$, 93
$\gamma_0, \ldots, \gamma_{k-m}$, 116, 122, 147, 153, 173, 182
$\gamma_0, \ldots, \gamma_{l-m}$, 189
λ, 23, 105, 110, 116, 122, 138, 142, 147, 153, 168, 173, 182
$\ln L$, 4
$\mathbb{E}(y)$, 2, 37, 49, 56, 105, 110, 111, 117, 123, 131, 138, 142, 148, 154, 168, 174, 183, 190, 197, 210
$\mathbf{A}_i$, 238
$\mathbf{R}_i(\boldsymbol{\alpha})$, 238
$\mathbf{y}_i$, 238
μ, 168, 173, 182, 189, 197
μ_{ij}, 237
ν, 189
ϕ, 2, 37, 168, 173, 182, 189, 197
π, 49, 56, 61, 116, 122, 147, 153

π^0, 52, 57, 62
π_0, 173, 189
π_1, 182, 189
ρ, 221, 222
σ^2, 1, 210, 221, 222, 281
$\sigma^2_{\tau_1}$, 296
$\sigma^2_{\tau_1}$, 281
$\sigma^2_{\tau_2}$, 281, 296
$\sigma^2_{u_1}$, 210, 248, 258, 281, 296
$\sigma^2_{u_2}$, 210, 248, 258, 281, 296
$\sigma_1, \ldots, \sigma_{p-1}$, 221
$\sigma_{\tau_1 \tau_2}$, 281, 296
$\sigma_{u_1 u_2}$, 210, 248, 258, 281, 296
τ, 189
τ_{1m}, 281, 290
τ_{2m}, 281, 290
θ, 2
$\tilde{y}$, 23
ε, 1
ε_{1jm}, 281
ε_{ij}, 210
$\widehat{r}$, 138, 143, 148, 153
$\widehat{\alpha}$, 37
$\widehat{\alpha}_1, \ldots, \widehat{\alpha}_{c-1}$, 73, 81, 87, 93
$\widehat{\beta}$, 37
$\widehat{\beta}_0, \ldots, \widehat{\beta}_k$, 3, 24, 37, 50, 57, 61, 106, 111, 138, 143, 168
$\widehat{\beta}_0, \ldots, \widehat{\beta}_m$, 117, 122, 148, 153, 174, 182, 189
$\widehat{\beta}_0, \ldots, \widehat{\beta}_{k+1}$, 210, 290
$\widehat{\beta}_1, \ldots, \widehat{\beta}_k$, 73, 81, 87
$\widehat{\beta}_{j1}, \ldots, \widehat{\beta}_{jk}$, 93
$\widehat{\gamma}_0, \ldots, \widehat{\gamma}_{k-m}$, 117, 122, 148, 153,

174, 182

$\widehat{\gamma}_0, \ldots, \widehat{\gamma}_{l-m}$, 189

λ, 106, 111, 117, 122, 138, 143, 148, 153

$\widehat{\mu}$, 168, 174, 182, 189

$\widehat{\nu}$, 189

$\widehat{\phi}$, 37, 174, 182

$\widehat{\pi}$, 50, 57, 61, 117, 122, 148, 153

$\widehat{\pi}_0$, 174, 189

$\widehat{\pi}_1$, 182, 189

$\widehat{\sigma}$, 3

$\widehat{\tau}$, 189

$\widehat{\widetilde{\zeta}}_0, \ldots, \widehat{\widetilde{\zeta}}_{k-l}$, 189

$\mathbb{E}(\tilde{y})$, 24

$\widehat{\mathbb{E}}(y)$, 3, 4, 37, 50, 282, 290

$\zeta_0, \ldots, \zeta_{k-l}$, 189

$c(\cdot)$, 2

$g(\cdot)$, 2, 237, 290

$h(\cdot)$, 2

r, 138, 142, 147, 153, 159

$t_1, \ldots, t_p$, 209, 237, 258, 281, 290

u_{1im}, 281, 290

u_{1i}, 210, 258

u_{2im}, 281, 290

u_{2i}, 210, 258

$x_1, \ldots, x_k$, 1, 2, 24, 37, 49, 56, 61, 72, 80, 86, 93, 105, 110, 116, 122, 138, 142, 147, 153, 168, 173, 182, 189

$x_1^0, \ldots, x_k^0$, 6, 25, 38, 52, 57, 62, 74, 81, 87, 94, 107, 111, 117, 123, 138, 143, 148, 154, 168, 174, 183, 190

$x_1^0, \ldots, x_k^0, t^0$, 211, 282, 290

$x_{1ijm}, \ldots, x_{kijm}, y_{ijm}$, 281, 290

$x_{1ij}, \ldots, x_{kij}, y_{ij}$, 210, 237, 258

$x_{i1}, x_{i2}, \ldots, x_{ik}, y_i$, 1

y, 1, 2, 37, 49, 56, 61, 72, 80, 86, 93, 105, 110, 116, 122, 138, 142, 147, 153, 168, 173, 182, 189

y^0, 6, 25, 38, 107, 111, 117, 123, 138, 143, 148, 154, 168, 174, 183, 190, 211, 282, 290

$\text{logit}\pi$, 50

Subject Index

z-score, 57, 81

AIC, *see* Akaike Information
    Criterion, 5
AICC, *see* Corrected Akaike
    Information Criterion, 5
Akaike Information Criterion
    (AIC), 5
asymptotic likelihood ratio test, *see*
    deviance test, 5
autoregressive covariance structure
    of errors, 221
autoregressive working correlation
    matrix, 238

Bayesian Information Criterion
    (BIC), 5
beta regression model, 167
BIC, *see* Bayesian Information
    Criterion, 5
binary logistic regression model, 49
binary variable, 49
Box-Cox transformation, 23

chance zero, 116
change in log-odds, 51
complementary log-log link
    function, 61
complementary log-log model, 61
compound symmetric covariance
    structure of errors, 222
Corrected Akaike Information
    Criterion (AICC), 5
count variable, 105

overdispersed, 137
covariance structure of errors
    autoregressive, 221
    compound symmetric, 222
    exchangeable, 222
    independent, 220
    spatial power, 221
    Toeplitz, 221
    unstructured, 220
covariate, *see* predictor variable, 1
cumulative complementary log-log
    model, 86
cumulative logit, 72
cumulative logit model, 72
cumulative probability, 72
cumulative probit model, 80

dependent variable, *see* response
    variable, 1
deviance, 5
deviance test, 5
dichotomous logistic regression
    model, *see* binary logistic
    regression model, 49
dichotomous variable, *see* binary
    variable, 49
difference in log-odds, 51
dummy variable, *see* indicator
    variable, 4

estimated mean response, 3, 4
estimated rate ratio, 87
exchangeable working correlation
    matrix, 239

exchangeable, *see* compound symmetric covariance structure of errors, 222

explanatory variable, *see* predictor variable, 1

exponential family of distributions, 2

extreme value distribution, *see* Gumbel distribution, 61

fitted mean response, *see* estimated mean response, 3

fixed-effects predictor, 209

gamma regression, 37

GEE, *see* Generalized Estimating Equations, 237, 265

general linear regression model, 1

Generalized Estimating Equations (GEE), 237, 265

generalized linear mixed model, *see* generalized random slope and intercept model, 257

generalized linear regression model, 2

generalized logit function, 92

generalized logit model, 93

generalized random slope and intercept model, 257

Gumbel distribution, 61

hierarchical regression model, 281, 289

hurdle negative binomial regression model, 153

hurdle Poisson regression model, 122

independent covariance structure of errors, 220

independent variable, *see* predictor variable, 1

independent working correlation matrix, 239

indicator variable, 4, 51, 57, 62, 73, 81, 87, 106

link function, 2
complementary log-log, 61
logit, 50
probit, 56

log-odds model, *see* binary logistic regression model, 50

logistic distribution, 50

logit link function, 50

logit of cumulative probability, *see* cumulative logit, 72

longitudinal data, 209

mean response, 2
estimated, 3

mixed-effects linear regression model, 209

model for clustered data, *see* hierarchical regression model, 281

multilevel, *see* hierarchical regression model, 281

multinomial logistic regression, 71

negative binomial regression, 137

nominal variable, 71

odds in favor of event, 50

one-inflated beta regression model, 181

ordinal variable, 71

outcome variable, *see* response variable, 1

overdispersed count variable, 137

percent change in odds, 51, 73

percent ratio in odds, 51

Poisson regression model, 105

polytomous logistic, 71

polytomous logistic regression, *see* multinomial logistic regression, 71

predicted response, 6, 25, 38
predictor variable, 1
probit index, *see* z-score, 57
probit link function, 56
probit regression model, 56

R fuction
    glm, 107
R function
    betareg, 169
    boxcox, 26
    clm, 74, 82
    gamlss, 176, 183, 191
    geeglm, 240, 265
    glm, 8, 9, 38, 52, 58, 63
    glm.nb, 139
    glmer, 259, 291
    hurdle, 124, 155
    lme, 213, 223
    lmer, 283
    logLik, 9
    multinom, 95
    pchisq, 9
    plotNormalHistogram, 8
    predict, 9
    print, 9
    relevel, 8
    shapiro.test, 8
    summary, 8
    vglm, 112, 143
    zeroinfl, 118, 149
R library
    geepack, 240, 265
    betareg, 169
    gamlss, 176, 183, 191
    lme4, 259, 283, 291
    MASS, 26, 139
    nlme, 213, 223
    nnet, 95
    ordinal, 74
    pscl, 118, 124, 149, 155
    rcompanion, 8

VGAM, 112, 143
random error, 1
random intercept, 210
random slope, 210
random slope and intercept model,
    210
random-effects term, 209
rate ratio, 87
regression
    beta, 167
    binary logistic, 49
    complementary log-log model,
        61
    cumulative complementary
        log-log, 86
    cumulative logit, 72
    cumulative probit, 80
    dichotomous logistic, 49
    gamma, 37
    general linear, 1
    Generalized Estimating
        Equations, 237, 265
    generalized linear, 2
    generalized linear mixed, 257
    generalized logit model, 93
    generalized random slope and
        intercept, 257
    hierarchical, 281, 289
    hurdle negative binomial, 153
    hurdle Poisson, 122
    mixed-effects linear, 209
    multinomial logistic, 71
    negative binomial, 137
    one-inflated beta, 181
    Poisson, 105
    probit, 56
    random slope and intercept,
        210
    zero-inflated beta, 173
    zero-inflated negative binomial,
        147
    zero-inflated Poisson, 116

zero-one-inflated beta, 189
zero-truncated negative
    binomial, 142
zero-truncated Poisson, 110
regression coefficient, 1
regressor, *see* predictor variable, 1
repeated measures, 209
response variable, 1
response, *see* response variable, 1

SAS procedure
    fmm, 111, 123, 154
    genmod, 7, 38, 52, 57, 62, 74,
        82, 88, 117, 148, 240, 265
    glimmix, 169, 258, 290
    logistic, 94
    mixed, 212, 223, 282
    nlmixed, 175, 183, 191
    transreg, 25
    univariate, 6
Schwartz Bayesian Information
    Criterion (BIC), 5
spatial power covariance structure
    of errors, 221
structural zero, 116

Toeplitz covariance structure of
    errors, 221
Toeplitz working correlation
    matrix, 238

unstructured covariance matrix of
    errors, 220
unstructured working correlation
    matrix, 238

variance function, 237
variate, *see* response variable, 1

working correlation matrix, 238
    autoregressive, 238
    exchangeable, 239
    independent, 239
    Toeplitz, 238
    unstructured, 238

zero-inflated beta regression model,
    173
zero-inflated negative binomial
    regression model, 147
zero-inflated Poisson regression
    model (ZIP), 116
zero-one-inflated beta regression
    model, 189
zero-truncated negative binomial
    regression model, 142
zero-truncated Poisson regression
    model, 110
ZINB, *see* zero-inflated negative
    binomial regression model,
    147
ZIP, *see* zero-inflated Poisson
    regression model, 116

Printed in the United States
by Baker & Taylor Publisher Services